TANNENBERG

The First Thirty Days in East Prussia

Tannenberg

The First Thirty Days in East Prussia

BY

MAJOR-GENERAL SIR EDMUND IRONSIDE,
K.C.B., C.M.G., D.S.O.

WITH MAPS

The Naval & Military Press Ltd

Published by

The Naval & Military Press Ltd
Unit 10 Ridgewood Industrial Park,
Uckfield, East Sussex,
TN22 5QE England

Tel: +44 (0) 1825 749494
Fax: +44 (0) 1825 765701

www.naval-military-press.com
www.military-genealogy.com
www.militarymaproom.com

PREFACE.

THE Campaign of Tannenberg is of peculiar interest to the British Army, because it shows what a small well-trained Force can achieve against numbers. Success does not lie in weight of numbers, but in the skilful employment in combination of all available arms. In the next war, success will go to that Army which possesses the best mechanical and destructive weapons, such as motor vehicles, aircraft, artillery, and tanks, and which can make the best use of them, thereby economising its man power.

All the factors for success lie within reach of the British Army. The Expeditionary Force of 1914 was undoubtedly superior, unit for unit, to the Forces of other Powers. It is our duty to see that our present Regular Army reaches the same standard. Our men have high *moral*, due to the active and independent life they lead, and they have many brilliant traditions behind them. They have a high standard of education, and that standard is being steadily improved. The more technical become the weapons of war, the more does the scale turn in our favour. We can make our men experts in the handling and employment of their weapons, whereas the short-service conscript will become merely a moderate adept.

The white population of the Empire belongs to a united race, joined together by racial sympathy and a common language, but our Imperial Power is slow to develop, and in the first instance our Regular Army must stand very much as did the Eighth German Army in East Prussia.

CONTENTS.

MAPS AND SKETCHES.

(At end of volume.)

TANNENBERG.

CHAPTER I.

The Initiation of the Campaign.

Map I.—Strategical Map.

1. Events Prior to the War.

The year 1914 found Europe in a state of armed suspense. Two great Political and Military Alliances stood facing one another—on the one hand Germany, Austria-Hungary, and Italy, and on the other France and Russia.

The Franco-Russian Alliance dated from 1892, a Military Convention between the two countries being concluded in 1893.

In the event of war with the opposing Alliance, the General Staffs of France and Russia were faced with a difficult problem in the co-ordination of the military operations of their two countries. Geographically, they were at a disadvantage in being widely separated, whereas the States of the German Alliance were centrally situated, cutting off all direct communication between France and Russia. Reduced to its simplest form, the military problem in the event of war between the two Alliances was roughly this: Could the French and Russians combine their operations so as to compel the German Alliance to disperse their forces in East and West and prevent them from establishing a preponderance in either theatre, and so crush them finally by superior man-power? Could the German Alliance hold off one of their opponents successfully while they defeated the other?

From 1893 to 1901 the operations to be undertaken by France and Russia were discussed by their General Staffs, and gradually embodied in a combined plan.

Both Staffs were agreed that the Central Alliance would certainly take the offensive at once. Germany's well-known efficiency in Mobilisation could only be turned to advantage by an immediate attack against one of her opponents. To act on the defensive was to await final defeat.

The rates of military concentration in France and Russia differed considerably owing to the distances to be covered in the latter country and the comparative inefficiency of the railways. France might expect to complete concentration in fifteen days, whereas Russia required twenty-eight days. The intervening period of thirteen days was therefore a critical one for the Franco-Russian Alliance. France, being the better prepared, would probably be attacked first, and it was essential to accelerate the entry of the Russian forces into the field as much as possible in order to prevent France from being overwhelmed before her slower Ally could make her weight felt.

The French General Staff wished as large a German force as possible retained in East Prussia, so that operations on the Western Front might have a chance of success. They pointed out that Germany, making use of her excellent railways, might carry out operations on interior lines upon an unexampled scale. In the event of a German success in the first great battle of the war on the Western Front, which might be expected between the fifteenth and eighteenth days of Mobilisation, Germany might then transfer a large force to the Eastern Front, and might still be in time to oppose Russia. It was argued, therefore, that it was to the advantage of France and Russia to assume a simultaneous offensive about the fifteenth day of Mobilisation. Germany would then be unable to draw the maximum advantage from her central position, and any attempt to annihilate one of her opponents before the arrival of the other would be prevented. The Russian General Staff pointed out that such an early advance on the part of Russia must inevitably mean a campaign undertaken with only a portion of the forces which would be available at a later date. There might even be a considerable danger in such an advance should Germany decide to attack Russia first, either with a view to supporting Austria or to protecting the favoured province of East Prussia. This contention was countered by the French General Staff, who pointed out that the French advance in the West would come with irresistible force through being less strongly opposed, and that a definite success in the West could be promised with certainty. This would no doubt lead to a cessation of any

German operations against Russia, where no quick decision was to be expected in any case.

The Russian Soviet Government has published an account of the Conferences which took place between the French and Russian General Staffs from 1901 to 1913. It is the only account at present available. The object of these Conferences was to scrutinise and revise the general plan of operations in the light of the changing political and military situation.

At the 1901 Conference, the French General Staff laid down two conditions which they considered essential for the success of the combined operations :—

(i) The diversion from the French Front of a portion of the German Army sufficient in size to permit of the success of a French offensive against Germany.

(ii) If the outcome of this offensive did not turn in favour of the French, then good news from the Russian Front must be at hand to react upon French Public Opinion and the *moral* of the French Army.

The first great battle of the war being expected about the fifteenth day of Mobilisation, the first condition would be fulfilled if, by the fourteenth day, such Russian forces were to advance against Germany as to permit of success against six to eight Corps. The second condition would be fulfilled if news of such an advance should arrive in France before the commencement of the first great battle on the Western Front.

The greatest stress was laid upon the importance of accelerating the Russian Mobilisation, and France was prepared to consider financial arrangements for the development of the necessary railways.

The disastrous war with Japan eliminated Russia temporarily from the European political scene. With the help of France, however, rapid strides were made in the reorganisation of the Russian Army, and by 1911 Russia once more became a factor in the military situation.

The 1911 Conference was attended by General Jilinsky, afterwards to command the Russian operations in East Prussia, and General Dubail, afterwards to command the First French Army in Lorraine.

In previous Conferences it had been decided that the main effort of Russia against Germany was to take the form of an offensive into East Prussia from the East, on the general line Kovno-Lomja. There now came an important modification, bearing in itself the seeds of future trouble. It was agreed that the centre of gravity of the Russian forces intended for action against Germany should be transferred farther South, with a view to two possible plans of action :—

(i) An advance in the general direction of Allenstein, should the Germans concentrate in East Prussia or attempt to advance from there against Warsaw.

(ii) An advance in the general direction of Berlin, should the Germans concentrate in the Thorn-Posen area or attempt to advance from there against Warsaw or Ivangorod.

Here was a complete new grouping of the Russian forces for action against Germany. Concentration was now to take place farther forward and in the Polish salient. A longer time would be required for the completion of the movement, unless the general speed of railway transportation could be raised, and all orders and time-tables would have to be redrafted. The Polish salient was exposed to attack from three sides, and the concentration would require adequate covering forces, in order to prevent its interruption by a sudden enemy thrust. The provision of these covering forces might even entail the alteration of the peace stations of the units concerned.

The French General Staff signified their readiness to advance on the twelfth day of Mobilisation with a force of 1,300,000 men. They admitted that a simultaneous advance at such a date was an unrealisable ideal, but urged the Russian General Staff to approximate as closely as possible to this date. For the moment it was the intention of the Russians to cross the frontier on the twentieth day. The French considered that this date should be advanced at all costs to the eighteenth day. The Russian General Staff pointed out that their Army was still in process of reorganisation, the scheme on foot not being due for completion until 1916. By that year Russia would be in a position to advance earlier than the eighteenth day with a force of 800,000 men. They pointed out that Austria-Hungary had improved her Mobilisation arrangements, and might be expected to take the offensive against Russia instead of standing on the defensive as had been expected before.

The 1912 Conference was attended by Generals

Joffre and Jilinsky. The Russian General Staff expressed fears with regard to the probable attitude of Sweden and Turkey in the event of war, and they also pointed out the increasing efficiency of the Austro-Hungarian Army. They now considered it necessary to leave certain troops in Finland and the Caucasus, which had not entered into their calculations heretofore.

The 1913 and final Conference took place in August. Generals Joffre and Jilinsky again represented their respective countries. German Mobilisation was being speeded up to a remarkable degree, and both Staffs were agreed that it was more than ever essential to advance against Germany as nearly simultaneously as possible. The French Army was ready to advance on the eleventh day of Mobilisation. The Russian General Staff engaged to advance as soon after the fifteenth day as possible, since on this date all would be ready with the exception of some of the transport and parks. The contingents for France and Russia were confirmed at 1,300,000 and 800,000 men respectively. The Russian forces were to be concentrated in such a manner that they could advance either into East Prussia or direct on Berlin, according to the initial German dispositions in the East.

The History of these Conferences shows clearly the correct view assumed by the French General Staff. They had striven all through to impress two facts upon their Ally—

(i) That Germany was the main enemy, and that the main force of the combined blow must be directed against her. Success against

Germany meant the collapse of the Central Alliance.

(ii) That the military operations on the Eastern Front must be subordinated to those on the Western Front, in order to make the most of the geographical situation of France and Russia.

These Conferences are interesting as showing the difficulties which have to be contended with by General Staffs in drawing up a combined plan for the action of Armies belonging to separate States. Great goodwill and tact are required on both sides. Moreover, as it is improbable that there will be an unified command at the commencement of any war in which two or more States are acting together, it is extremely important that the Commanders of the separate Armies should have a thorough knowledge of, and be in complete sympathy with, the combined plan as worked out beforehand. General Joffre became Generalissimo of the French forces, whereas General Jilinsky was only responsible for the conduct of operations in East Prussia, and did not direct the general timing of the Russian operations.

2. Dates of Mobilisation.

The dates of the various Mobilisations and Declarations of war were as follows :—

28th June. Assassination of the Archduke Franz Ferdinand at Serajevo.

5th July. Council at Potsdam.

23rd July. Austrian Ultimatum to Serbia.

25th July. Serbia orders General Mobilisation.

26th July. Austria orders Mobilisation of those forces which would naturally be employed against Serbia in the event of war.

28th July. Austria declares war against Serbia.

31st July. Russia orders General Mobilisation.
Austria, on receipt of this information, also orders General Mobilisation.
Turkey orders Mobilisation.

1st Aug. France orders General Mobilisation.
Germany declares war on Russia.

2nd Aug. German Ultimatum to Belgium.
Hostilities commence between France and Germany.

3rd Aug. Belgium refuses German Ultimatum.
Germany declares war on France.
Great Britain orders General Mobilisation.
Italy declares her Neutrality.

4th Aug. Great Britain declares war on Germany.
Germany declares war on Belgium.

6th Aug. Austria-Hungary declares war on Russia.

3. General Political Situation at the Outbreak of War.

The first event of importance was the default of Italy from the German group. This eased considerably the situation of France and Russia. They now found themselves allied with Serbia against Germany and Austria-Hungary alone. Diplomacy had so far served them well. The next event was the entry of Great Britain into the struggle on the side of France and Russia. France's coast-line was secured, and a well-trained force added to the credit of the French Army in the West. In point of man-power, therefore, the Franco-Russian group was distinctly superior to its opponents.

The first few days of the war showed clearly that the Germans were going to carry out their main offensive against France, leaving Austria and a small residue of the German forces to deal with Russia.

The French General Staff had foreseen correctly the nature of the German plan.

4. Theatre of Operations.

(i) *Historical.*

No portion of the Earth's surface has been more fought over than East Prussia. Its history, like that of so many border states between East and West, has been one continual struggle to stem the advance of barbarian hordes. The fear of danger from the East has always been very real. The inhabitants of the Dukedom of Masovien were undoubtedly of Slav origin. Their entry into the territory lying between the Memel and the Vistula is shrouded in mystery. They were probably the pioneers of the movement from East to West which brought the Finns and Hungarians into Europe. In 1226 the

Duke of Masovien, finding himself unable to withstand the inroads of the Poles, Lithuanians, and Tartars, turned for help to the Order of Teutonic Knights. The Grand Master of those days, von Salza, arrived from Venice with the full approval of the Pope, and the stern rule thus instituted lasted nearly two centuries. By the introduction of many German settlers, the upper ruling class became Teutonic, the Masurens or Prussians willingly subjecting themselves to their military masters in return for the measure of protection afforded to them.

In the year 1410, however, at the Battle of Tannenberg, the Teutonic Knights were utterly routed by their enemies the Poles and Lithuanians, and the Teutonic Order ceased to exist. The great cities of Danzig, Elbing, Marienburg, and Thorn passed into the hands of the Poles, and the Peace of Krakau in 1525 converted East Prussia into a dependency of Poland. It was not for another century that the Province again passed under Teuton sway as a result of the successful campaigns of the Great Elector.

The Russo-Swedish wars and the campaigns of Napoleon took their toll of the country in money and lives, until the retreat from Moscow in 1812. Names such as Gross-Jägersdorf, Pultusk, Eylau, Ostrolenka, Heilsberg, Friedland, Tilsit, and Königsberg bear witness to the trials of the country.

During the century of peace preceding the Great War of 1914, East Prussia had developed steadily under the hard but just rule of the German upper class. So Germanised did the country become, that notwithstanding a large Polish and Masurisch-speaking peasantry, the people of East Prussia had come to be regarded as the backbone of Germany.

In the days of 1914 the fear of Russia was no less great than in earlier days. One had only to cross the Frontier into Old Russia to find a different standard of living, the neat well-kept homesteads changing to dirty unkempt hovels. Every harvest-time shoals of Poles and Russians used to swarm across for the reaping, and the East Prussian farmer had ample evidence of the less civilised state of his Eastern neighbours, fearing them all the more in times of violence. The word "Cossack" conjured up a bogey even in the hearts of the well-educated. Russian man-power had become a sort of fable in the military world. Immense, unlimited, irresistible—such were words employed. If we in England awaited the advance of the Russian "steam-roller," how much more menacing must it have appeared to the German peasant farmer?

(ii) *Geographical.*

Prussia was the home of the ruling class in Germany, and peculiarly dear to their Hohenzollern leaders. It was therefore a delicate spot in the armour of German defence.

Geographically, East Prussia points forward into the heart of Slavdom, and is exposed to invasion from both East and South, a most difficult problem of defence. The configuration of the Frontier offered an excellent opportunity for a concentric advance by the Russians against any German forces left to defend this portion of their territory.

The terrain, however, over which a Russian advance must take place was exceedingly difficult. Lakes, swamps, and forests restrict the movement of troops, and make the country peculiarly adaptable for defence.

Roads and railways on the Russian side of the Frontier were few and bad. The Russian forces were therefore handicapped in their rear services before they reached Prussian territory. Roads in East Prussia were numerous and good, securing mobility to the German forces. Should the Russian forces reach East Prussia, they had not the necessary motor transport to take advantage of the good roads, nor were they likely to be able to capture sufficient rolling-stock to operate the railways, which were of different gauge to their own.

Taking the East Prussian Front from North to South, we can divide it as follows :—

(*a*) *Königsberg Area.*

A more or less fortified area extending over the Königsberg peninsula.

From Labiau along the R. Deime to Tapiau the river offers an excellent line of defence. It is navigable for small river steamers and not fordable. The banks are always marshy. There are road-bridges at Labiau and Tapiau and a bridge just East of Labiau for the Tilsit-Königsberg railway, but no others. The Western bank has a considerable advantage in height and excellent observation of all the exits from the forest, which runs close up to the Eastern bank.

From Tapiau to Brandenburg and the fortified port of Pillau, small field-works had been traced out, forming the framework of a good line of defence, capable of being quickly strengthened.

In the centre lay the fortress of Königsberg with its obsolete inner works, strong against anything but heavy artillery. Access to Königsberg was available to ships by the canal from Pillau. Thus the Königsberg area formed an enclave of a depth of 19 miles from North to South. It could hardly be forced, in the absence of heavy artillery, without considerable loss of time and heavy casualties. It could be reinforced at any time by sea, and formed a formidable flanking position from which to menace any force passing it to the South.

(*b*) *The Insterburg Gap.*

A gap of 43 miles between Tapiau and Angerburg, down the centre of which runs the Insterburg-Allenstein railway. The country is on the whole flat, the general undulation never assuming a height of more than a hundred feet. There are several forest areas, notably those of Rominten near the Frontier, and the district just South of Wehlau. Although these forests are carefully tended and have many paths, they offer a considerable obstacle to the advance of large bodies of troops. The remainder of the country is open, large farms and small villages, each with its clump of trees, being dotted about the fields of corn and fodder, none of which are enclosed. The country is suitable for the movement of all arms, except at a few points where there are marshy bottoms. These are easily distinguishable from the presence of small heaps of black peat which is cut from the swamps.

The field of view is often extensive, and apart from the forest areas, the country, especially in summer, when the dust is excessive, offers an excellent terrain for observation from aircraft.

The R. Angerapp, running North and South from Angerburg, blocks this gap. It forms an excellent line of defence, the right flank resting against the impassable Lake Mauer. Though the river is fordable, it offers an important military obstacle. The banks are in places marshy, and the Eastern is commanded from the Western. The Germans had reconnoitred this position thoroughly, and it could be quickly occupied and strengthened.

(*c*) *The Masurian Lakes position.*

A line of continuous lakes, stretching from Angerburg to the Russian Frontier South of Johannisburg, a distance of 50 miles. The narrow gaps between the lakes were known to have been thoroughly reconnoitred, and were being busily strengthened by the Germans. Fort Boyen, guarding the Lötzen Gap, though obsolete, could only be reduced by heavy artillery, owing to the impossibility of approach across an absolutely open glacis. The Lakes are everywhere navigable, and there were many small steam-boats for general traffic. These had either been turned into miniature gun-boats or had been removed by the Germans. The shores of the Lakes often rise to a hundred feet in height, the forest and marshy reedlands running close up to the water.

The line of Lakes thus presented an impregnable front, capable of defence by very few troops, and offering an excellent screen, from behind which the German troops could operate to either flank.

(*d*) *The Johannisburg-Soldau Front.*

A distance of 75 miles along the Frontier running East and West. On the Russian side, the country North of Ostrolenka was a sandy forest waste, intersected with marshy streams and inhabited by a few Polish peasants. The forest areas were badly tended, and the roads mere tracks.

This area had been deliberately left a waste as a defensive measure against Germany, and constituted a difficult military obstacle.

On the German side the roads were good, but ceased as if by magic at the Frontier. The country is generally more cultivated and more closely inhabited. It is wooded and close, with the exception of the area immediately West of Neidenburg, but there are no enclosures anywhere.

The most important forest area is that of Johannisburg, where the Germans had constructed a line of blockhouses to throw back a flank to the line of Lakes.

Between the line Willenberg-Neidenburg and the line Allenstein-Osterode the country is difficult, and favours the defence. Everywhere there are sunken lakes, their chief feature being their elongated nature from North to South. Thus lateral communications for a force advancing Northwards were difficult.

A Russian advance into East Prussia was therefore faced by many natural difficulties, and the area available for manœuvre along the whole front considerably restricted by important obstacles.

CHAPTER II.

The Forces Engaged.

1. Composition of Higher Units.

(i) *Russian.*

An *Infantry Division* was composed of 2 Infantry Brigades, each of 2 Regiments, each of 4 Battalions, thus 16 Battalions in all. The Artillery consisted of 1 Artillery Brigade of 6 Batteries, each of 8 guns, in all 48 guns.

The number of all arms was about 20,000, and the *Fighting Strength* 14,000 *rifles and* 48 *guns.*

A *Rifle Brigade* was composed of 4 Rifle Regiments, each of 2 Battalions, thus 8 Battalions in all. The Artillery consisted of 1 Rifle Artillery Division of 3 Batteries, each of 8 guns, in all 24 guns.

The number of all arms was about 9,500, and the *Fighting Strength* 7,000 *rifles and* 24 *guns.*

A *Cavalry Division* was composed of 2 Brigades, each of 2 Regiments, each of 6 Squadrons.

The number of all arms was about 4,500, and *Fighting Strength* 3,500 *sabres and* 12 *guns.*

(ii) *German.*

An *Infantry Division* was composed of 2 Infantry Brigades, each of 2 Regiments, each of 3 Battalions, thus 12 Battalions in all. The Artillery consisted of 1 Artillery Brigade of 2 Regiments, each of 2 Abteilungen, each of 3 Batteries, each of 6 guns, in all 72 guns.

The number of all arms was about 17,500, and the *Fighting Strength* 12,000 *rifles and* 72 *guns.*

A *Cavalry Division* was composed of 3 Brigades, each of 2 Regiments, each of 4 Squadrons.

The number of all arms was about 5,000, and the *Fighting Strength* 3,600 *sabres and* 12 *guns.*

2. Comparison of the Two Opposing Armies.

(i) *Personnel.*

The educated class in Russia was small. In the Army, at a liberal estimate, it may be taken that only 50 per cent were able to read and write. In the German Army, the percentage of illiterates was negligible.

The difficulty of finding Officers and Non-Commissioned Officers for Russia's vast Army had never been solved by the Russian General Staff. Of the serving N.C.O.'s 75 per cent were conscripts, having therefore insufficient service to be efficient. The Reserve for both Officers and N.C.O.'s in Russia was woefully deficient. In Germany, the Reserve of Officers and N.C.O.'s was proportionately the highest in Europe.

Russian units were composed of two-thirds Russians and one-third subject races, such as Letts, Balts, and Poles. On Mobilisation, units were filled up by local reservists. German units, on the other hand, were strictly territorial, the small number of Polish subjects being distributed amongst the Regiments, so that their presence was not noticeable.

As a result, far more initiative and independence was to be expected from the German soldier as compared with the Russian. German subordinate leading was superior throughout, and the Russian units, for the purpose of economy in leaders, were bigger and more clumsy than German units. The esprit de corps of German units was distinctly higher than that of Russian units.

(ii) *Armament.*

The Germans were superior to the Russians in Artillery, especially so with regard to Heavy Artillery, the proportion of Guns to Rifles in the two Armies being :—

Russian Army . .	1 Gun to 292 Rifles.
German Army . .	1 Gun to 166 Rifles.

In Heavy Artillery, the superiority in favour of the Germans was 10 to 1.

A German Infantry Division, with 12,000 rifles and 72 guns, was a much stronger fighting instrument than a Russian Infantry Division with 14,000 rifles and 48 guns.

The machine-guns on both sides amounted to two per Battalion.

In Russia, reserves of all armament were small—an important fact to remember when taken in conjunction with Russia's small industrial resources and isolated position.

(iii) *Transport.*

At the outbreak of war Russia possessed only 679 motor vehicles, while there were 475 civilian motor vehicles in the country which could be requisitioned. The Germans, on the other hand, were well supplied. The Russians had to depend wholly upon horse transport in front of railhead, the mass of horses required considerably complicating the supply situation. The German Army was thus immeasurably more mobile than the Russian Army. General Knox, in his book, compares the Russian Army to a heavy-weight, muscle-bound prize-fighter, who, because of his enormous bulk, lacked activity and quickness, and would therefore be at the mercy of a lighter but more wiry and intelligent opponent.

(iv) *Aircraft.*

Aircraft did not play a great part in the campaign in East Prussia. The Russians possessed 320 aeroplanes at the outset of hostilities, and these were employed almost wholly on the Austrian front. The Germans retained few machines on their Eastern front, and their air reconnaissances failed on more than one occasion.

(v) *Training.*

The general standard of training in the Russian Army was inferior to that in the German Army, chiefly owing to the lack of method inherent in the Russian character. Great strides had, however, been made since the days of the Manchurian War, especially in the Artillery. The Cavalry appears to have failed to keep up with the rest of the Army, and was the most deficient in training of all the arms.

(vi) *Technical Equipment.*

In the Russian Army the most serious deficiency lay in the lack of telephone, wireless, and telegraph personnel and equipment. Communication suffered throughout the Campaign, thus still further immobilising the Army. On the other hand, the Germans were well supplied with signal equipment.

3. Available Strengths.

A Field-State of the First Russian Army, dated the 14th August, gives its strength as follows :—

6 First-Line Infantry Divisions.
1 First-Line Rifle Brigade.
7 Second-Line Infantry Divisions.
5½ First-Line Cavalry Divisions.

A Memorandum prepared by the Chief of Staff

of the Second Russian Army, dated the 14th August, gives its strength as follows :—

12 First-Line Infantry Divisions.
1 First-Line Rifle Brigade.
4 Second-Line Infantry Divisions.
3 First-Line Cavalry Divisions.

The strength of the Eighth German Army on the 14th August is given as follows :—

6 First-Line Infantry Divisions.
5 Second-Line Infantry Divisions.
1 First-Line Cavalry Division.

The comparison of the opposing forces therefore works out as follows :—

	Infantry Divisions. First-Line.	Infantry Divisions. Second-Line.	Cavalry Divisions.
Germans .	6	5	1
Russians .	19	11	$8\frac{1}{2}$

This meant a superiority of 2·7 to 1 in Infantry and 8·5 to 1 in Cavalry in favour of the Russians.

4. Commanders.

The Russian Commanders had all risen to high rank at a comparatively youthful age. They had had opportunities of commanding large units in peace. The majority had gained recent war experience in the Manchurian War, whereas the German Commanders had no recent war experience. By their system of promotion to high rank, the Russians had narrowed the talent available. The great majority of vacancies for Regimental Commanders in the Infantry and Cavalry of the Line were given to Officers of the General Staff or Guard Corps to the exclusion of all others. Ambitious Officers who could pass the examination went to the Staff College very young, before they had had sufficient Regimental experience. After three years' study they were appointed to the General Staff. General Knox quotes a Russian newspaper article giving the career of the average General Staff Officer. They left the Staff College usually with six to eight years' service. They were then supposed to command a company or squadron, but seldom did in actual fact. The next four years were spent in a subordinate position on the Staff of a Division or Corps, and the young officer was out of touch with troops. Six years after leaving the Staff College, that is to say when he had twelve to fourteen years service, he became a Lieutenant-Colonel. He was then transferred to the Staff of his District or to Army Headquarters, where his work remained the same. He never decided anything, and never expressed an opinion of his own, but spent his time in collating the opinions of others.

Such a career ill fitted them to compete with the German Commanders, who had been trained to hard work and energetic action.

The German Command and Staff had been created by years of profound study of war in all its bearings. Study of Military History replaced, as far as it could, the actual experience of war, and the German machine worked with mechanical precision.

The greatest weakness of the Russian Command and Staff lay in their ignorance of organisation and administration, which were perhaps the strongest points of the Germans.

A brief summary of the careers of the leading Commanders on both sides in the East Prussian Campaign is given below.

(i) *Russian.*

Jilinsky.

Commenced his service in the Chevalier Guard Regiment. He served on the Staff of the Viceroy Alexieff in the Japanese War. After the war of 1904 he commanded a Cavalry Division in Poland. He was appointed Chief of the General Staff in 1910, holding the post till 1913. During this period he took part in all the negotiations with the French General Staff. He was appointed to command the Warsaw Military District in 1913, and on the outbreak of war was given the Command of the N.W. Front. He was 60 years of age at the outbreak of war. He appears to have been more of an office soldier than a leader in the field. He was also unpopular with his immediate subordinates, and did not possess the confidence of the Army as a whole.

Paul Rennenkampf.

Rennenkampf was 60 years of age at the outbreak of war. He passed out of the Academy of the General Staff at the age of 28. At the age of 41 he received a Regiment, and promotion to Major-General at 46. In 1900 he commanded a column against the Chinese, and a Cossack Cavalry Division in 1904 against the Japanese. In the latter war he finally commanded forces of all arms, particularly distinguishing himself for his energy and dash. From 1905 to 1913 he commanded a Corps, when he was appointed to the command of the Vilna Military District.

Opinion in the Russian Army was divided as to whether he or Samsonoff was the finer commander. Both men and officers had confidence in his powers of leadership. He appears to have deteriorated much during the period of peace soldiering since 1904. His Baltish origin led to many rumours as to his loyalty in 1915, when he finally lost his head and deserted his Army. He was openly accused of traitorous conduct, and was dismissed in disgrace from the Army. He was killed by the Bolsheviks in 1918.

Alexander Samsonoff.

Samsonoff was 55 years of age at the commencement of the War. He passed out of the Academy of the General Staff at the age of 25. He served in the war of 1877 against Turkey. After commanding a Cavalry School he was promoted to Major-General at the age of 43. In the Manchurian War he successively commanded a Cavalry Brigade and a Cavalry Division. In 1909 he became Governor of Turkestan. He was on sick leave in the Caucasus when war broke out, being then appointed to the command of the Second Army. He was much liked and respected by all ranks, and there was a general confidence in his military capacity. Both physically and mentally, however, Samsonoff appears to have degenerated from the days of the Manchurian War, and semi-military employment had unsuited him for leading large forces in the field.

(ii) *German.*

Max von Prittwitz und Gaffron.

Von Prittwitz was 66 years of age when war broke out. He had joined in April 1866 at the age of 17, and took part in the campaigns against Austria and France. In the latter, he gained the Iron Cross of the 2nd Class. He spent three years at the Kriegsakademie, from '73 to '76, thus passing out at the age of 27. For an Infantry officer his promotion was comparatively rapid. Colonel at 44, and Major-General at 47. At the outbreak of war he had been a General officer for upwards of 17 years in regular employment, and was at the moment an Army Inspector with the rank of Generaloberst.

Paul von Beneckendorff und Hindenburg.

Hindenburg was 67 years of age when war broke out. He was thus a year older than Prittwitz, who was his cousin. He had joined in April 1866, and took part in the campaigns of '66 and '70. He was present at the battles of Trautenau, Königgratz, St Privat, and Sedan. He gained the Iron Cross of the 2nd Class after Sedan. He passed the same three years as Prittwitz at the Kriegsakademie, passing out at the age of 28. An Infantry officer also, his promotion was slower. Colonel at 47, and Major-General at 50. In 1911 he retired at his own request. When recalled to active service he had therefore been out of employment for three years.

A man of large and ponderous figure, with the broad heavy face of the Prussian, his very stolidness inspired a sense of security. He supplied the necessary bulwark behind which the nimble brain of Ludendorff could work out the details of the operations they directed together.

August von Mackensen.

Mackensen was 65 years of age at the outbreak of war. Mackensen did not come of a military family, and it was not till 1899 that he was granted the ennobling "von." He joined the Hussars of the Life Guard as a one-year Volunteer while a student at a University in 1869. He took part in the campaign against France, becoming a Lieutenant of the Reserve. After returning to the University, he finally entered the Hussars of the Life Guard in 1873 at the age of 24, considerably older than the average. He never graduated at the Kriegsakademie, but was appointed to the General Staff as Captain. He owed his promotion to the fact that he was History Instructor to the Kaiser during his Military training, and afterwards the Commanding Officer of the Crown Prince in the Death's Head Hussars. Colonel at 48, and Major-General at 49, he had received especially rapid promotion. The war found him at the head of the XVIIth Corps.

A fine handsome man with grey hair, he presented a picture of the typical cavalryman of history.

At the Armistice he was interned by the French at Neusatz, being finally liberated in December 1919.

Erich Ludendorff.

Ludendorff was 49 years of age at the commencement of the war. Like Hindenburg, he hailed from near Posen. His promotion was also far from rapid.

After long service on the Staff, in 1913 he was given the command of a Regiment, at the age of 48. He was then reappointed to the Staff at the outbreak of war. His fine work at Liège resulted in his being chosen at the moment of crisis to be Chief of Staff to Hindenburg.

Ludendorff represented the German General Staff with all its merits and failings. He had an accurate painstaking brain with a capacity for immense hard work. With no other pastime than his profession, all his abundant energy was devoted to study and work. He had immense belief in himself and in the invincibility of the German Army. He had a contempt for anything civilian, and desired nothing better than military autocracy, where there would be no backsliding to weakness. There is little doubt that it was the brain of Ludendorff which produced the various operations of the German Army from the date he came to Supreme Headquarters. The solid cloak of such a man as Hindenburg was the best setting for a man like Ludendorff. He never actually held command in war, but took his responsibility in the name of another.

CHAPTER III.

The Opening of the Campaign.

Map I.—Strategical Map.

1. General Dispositions.

(i) *French Forces.*

The French Mobilisation and Concentration proceeded according to plan. On the 14th August the First and Second Armies attacked in Lorraine, while the VIIth Corps and a Cavalry Division advanced on Muhlhausen. In the first few days these attacks succeeded in advancing a considerable distance, but were finally brought to a standstill about the 24th August—firstly, to find the necessary troops to meet the German advance through Belgium; and secondly, because they had run against very strong and semi-permanent defences, which could not be reduced without deliberate operations.

It will thus be seen that the French General Staff had fulfilled their engagements to the letter.

(ii) *German Forces.*

The German General Staff had staked their all upon the advance through Belgium, and their superior Mobilisation arrangements had enabled them to commence their advance in front of their opponents. The Belgian frontier was actually violated on the 3rd August.

In Lorraine and Alsace the Germans had succeeded in their objective, checking the French offensive.

In East Prussia, General von Prittwitz had located his available troops, Eleven Infantry Divisions and a Cavalry Division, as follows :—

(*a*) Three Infantry Divisions (Two First-Line and One Second-Line) and One Cavalry Division, based on Insterburg, facing Eastwards towards Vilna.

(*b*) Three Infantry Divisions (Two First-Line and One Second-Line), based on Allenstein, facing Southwards towards Poland.

(*c*) One weak Second-Line Brigade holding the Lake position.

The remainder of his forces were located at various points to the West of the Lake position, from whence they could reinforce either of the two groups at Insterburg and Allenstein.

(iii) *Russian Forces.*

The Russians had ordered their Concentration in accordance with a plan drawn up in 1910.

The detail of this plan was as follows :—

Case A (Austria).

If Russia were opposed by the main Austrian Army and the residue of the German Army left over from operations in France.

Case G (*Germania*).

If Russia were opposed by the main Armies of both Austria and Germany.

All movements were identical for both cases up to the seventh day of Mobilisation. A decision had then to be taken as to which Case was to be followed.

In Case A, the First and Second Armies opposed the Germans, while the Third, Fourth, and Fifth Armies opposed the Austrians.

In Case G, the First, Second, and Fourth Armies opposed the Germans, while the Third and Fifth Armies opposed the Austrians.

The Sixth and Seventh Armies were to be in Reserve.

In Case A, the positions of concentration of the First, Second, and Fourth Armies were as follows :—

First Army.

First-Line Troops.	Guard Corps.	Kovno.
	I^st Corps.	Orany.
	IIIrd Corps.	Orany, Olita, and Simno.
	IVth Corps.	Druskienki.
	5th Rifle Brigade.	Suwalki.
Second-Line Troops.	53rd Division.	Poneviej.
	51st Division.	Radziwilishki.
	54th Division.	Kovno.
	73rd Division.	Kovno.
	57th Division.	Olita.
	68th Division.	Riga.
	72nd Division.	Orany.

In all, Eight First-Line Divisions, a Rifle Brigade, and Seven Second-Line Divisions.

In Case G, the 53rd, 54th, and 68th Divisions joined the Fourth Army, being replaced by the 80th and 82nd Divisions from the Fourth Army.

There was therefore a reduction of One Second-Line Division in Case G.

Second Army.

First-Line Troops.	IInd Corps.	Augustow.
	VIth Corps.	Lomja.
	XIIIth Corps.	Sokolka.
	XVth Corps.	Zambrowo.
	XXIIIrd Corps.	Bialystok, Sokolka, Volkovisk.
	1st Rifle Brigade.	Warsaw.
Second-Line Troops.	59th Division.	Siedlets.
	76th Division.	Grodno.
	77th Division.	Warsaw.
	79th Division.	Malkin.

In all, Ten First-Line Divisions, a Rifle Brigade, and Four Second-Line Divisions.

In Case G, the Grenadier and XVIIth Corps joined from the Fourth and Fifth Armies respectively.

There was therefore an increase of Four First-Line Divisions in Case G.

Fourth Army.

First-Line Troops.	Grenadier Corps.	Kholm.
	XIVth Corps.	Lyublin.
	XVIth Corps.	Lyublin.
	XXth Corps.	Lyublin.
Second-Line Troops.	80th Division.	Luckov.
	82nd Division.	Luckov.
	83rd Division.	Ivangorod.

In all, Eight First-Line Divisions and Three Second-Line Divisions.

In Case G, the whole of the Fourth Army was to concentrate on the right flank, North of the First Army, instead of against the Austrians.

The 80th and 82nd Divisions joined the First Army, being replaced by the 53rd, 54th, and 68th Divisions from the First Army. The Grenadier Corps joined the Second Army and the XIVth Corps the Fifth Army.

There was therefore a reduction of Four First-Line Divisions and an increase of One Second-Line Division in Case G.

The Objectives of the North-West Front, as the Command directing the operations against the Germans was called, were laid down as follows :—

Case A.

Headquarters to be established at Volkovisk.

General Objective of the Front : To defeat the German Forces left in East Prussia, and to occupy that Province. Operations to be arranged with a view to creating a favourable position for an advance deeper into Germany.

First Army—Headquarters, Vilna.

Initial Objective of the Army : To reconnoitre the enemy positions from Polangen (on the Baltic N. of Memel) to Lyck inclusive, and to observe the Baltic Coast within the Army area. To cover the Mobilisation and general Concentration. To make preparations for an Advance, paying particular attention to turning the Masurian Lakes to the North.

Second Army—Headquarters, Volkovisk.

Initial Objective of the Army: To reconnoitre the enemy positions in East Prussia to the West of Lyck, and to observe the German Frontier in the Army area. To cover the Mobilisation and general Concentration. To make preparations for an Advance, paying particular attention to turning the Masurian Lakes to the West. The Bialystok-Grodno area to be specially protected.

Case G.

Headquarters to be established at Volkovisk.

General Objective of the Front: To advance against those German Forces threatening Russia from the direction of East Prussia. The enemy to be held on all other Fronts.

Fourth Army—Headquarters, Szweksznie.

The Army to be concentrated in the Riga-Shavli area.

Initial Objective of the Army: To reconnoitre the enemy positions from Polangen to Schmalleningken inclusive; to observe the Baltic Coast in the Army area; to cover the Mobilisation and general Concentration; to make preparations for an Advance on the Front Tilsit-Insterburg, paying particular attention to the protection of the Right Flank of the Front.

The Initial Objectives of the First and Second Armies were to remain the same as in Case A, with the following alterations :—

(i) The First Army to be concentrated on the Front Schmalleningken-Lyck inclusive.

(ii) The Second Army to be concentrated farther to the West near Ostrolenka, with a view to preliminary operations by the IInd and VIth Corps, operating under one Commander, but directed by the Commander of the Second Army.

(iii) The First and Second Armies to be prepared to carry out operations in close co-operation with one another.

2. Comments upon the German and Russian Dispositions.

German.

The German Dispositions appear excellent. The Lake Position is lightly held, while two strong groups are thrown out in the direction of probable danger. The excellent East Prussian Railway system admits of reinforcement of either group, quickly and without enemy interruption. The remaining forces are disposed centrally on the railway, ready to move either North or South of the Lakes. These Dispositions are a good example of a concentration for a Strategic Defensive.

Russian.

Russia disposed of a total of 114 Divisions. She was deploying 73 Divisions in her initial operations, while holding 13 in Reserve. The remaining 28 Divisions came from her distant territories, and could not be counted upon in the first few weeks of the war.

The actual detail of the 1910 Plan works out as follows :—

Case A.

30 Russian Divisions against a possible 16-25 German Divisions.

43 Russian Divisions against a possible 43-47 Austrian Divisions.

Case G.

41 Russian Divisions against a possible 67 German Divisions.

32 Russian Divisions against a possible 43-47 Austrian Divisions.

The Sixth Army was to contain 7 Divisions.

The Seventh Army was to contain 6 Divisions.

The Plan, therefore, gives no indication of a desire to concentrate a superiority of force against one opponent, while holding the other with a minimum of force. When compared with the bold German plan of concentrating everything against France, it appears weak. The Plan is essentially defensive in the early stages of the proposed campaign. Only in Case G would the Russian contingent reach the 800,000 men agreed upon, and, moreover, there did not appear to be a likelihood of a great many of the Divisions crossing the frontier shortly after the fifteenth day of Mobilisation, owing to the distances they would have to cover from their areas of concentration.

3. Events succeeding Russian Mobilisation.

On the 1st August the Russian Military Attaché in Paris wired to the effect that the French General Staff were of opinion that the Russian advance should take place in the direction of Posen. They declared that it was certain that Germany was directing her main forces against France. The Russian General Staff was further asked to use their influence with the Serbians to urge them to advance at once against Austria. By engaging the attention of this country, more Russian troops would be set free for the advance against Germany.

On the 3rd August came the news of the violation of Belgian territory by the Germans. There could now be no longer any doubt that the main German effort was being made against France.

After the Conference of 1913, the office of Chief of the General Staff in Russia had passed from General Jilinsky to General Januschkevitch. The latter, by virtue of his office, became Chief of Staff to the Russian Generalissimo upon the outbreak of war. The programme of 1910 was in full swing. No immediate offensive against Germany had been arranged, and there was no possibility of an early advance against Posen. The difficult decision as to what was to be done lay in the hands of Januschkevitch. There was yet time before the seventh day of Mobilisation (6th August) to order the heavier concentration against Germany. This heavier concentration had only been envisaged in case Germany developed her main effort against Russia. This she evidently did not intend to do. Further, the heavier concentration entailed a dislocation, if not an abandonment, of the offensive against Austria, since the Fourth Army would be taken from the right flank of the Austrian front. Such a move tended to bring the Russian centre of gravity to the North on the German front and away from the Polish salient where the French desired it. Januschkevitch decided to allow the offensive against Austria to continue according to plan, and to improvise a fresh campaign against Germany.

On the 6th August he wired to Jilinsky :—

"It is essential that we prepare for an energetic offensive against Germany at the earliest possible moment, in order to ease the situation of the French, but, of course, only when sufficient strength has been made available. In view of the fact that Germany has directed her main forces against France, and that it is essential to support our Ally, the following measure will be carried out :—

(i) The Guard and Ist Corps will be withdrawn from the First Army, and will be directed upon Warsaw, where they should arrive between the 12th and 22nd days of Mobilisation (11th to 21st August).

(ii) The XXth Corps (from the Fourth Army) will join the First Army.

(iii) The XVIIIth Corps (from the Sixth Army) will move to Warsaw.

(iv) The units in Warsaw will form the Advanced Guard of a new Army for action against Germany. They will be under the orders of the Commander of the North-West Front."

Januschkevitch was attempting to shift the centre of gravity of his offensive against Germany as far South as possible, in accordance with the plan agreed upon subsequent to 1910. The train movement of the XXth Corps involved no difficulty, since it was the left Corps of the Fourth Army, and merely followed its programme according to Case G. The passing of the Guard and I[st] Corps right across the rear of the communications of the First and Second Armies must, however, have tried the railways a good deal, and probably delayed the rear units of these two Armies considerably. The new Army at Warsaw could not be assembled until the twenty-seventh day of Mobilisation (26th August) at the earliest, and therefore could not be expected to make an immediate advance against Posen. Though the plan did not dislocate the arrangements for the offensive against Austria, it reduced one of the Armies on this front by one Corps.

Jilinsky, who was, of course, well aware of Russia's engagements as regards timing and numbers, then sent in a plan proposing an advance of the First and Second Armies on the 13th August. This date would have coincided with the French advance in Lorraine, but it must be remembered that the

Russian Corps had farther to go to reach the frontier than had the French in the West.

Jilinsky's plan was in substance as follows :—

First Army to advance on the line Insterburg-Angerburg.

Second Army to advance on the line Rastenburg-Rothfliess.

IInd Corps to advance from Grodno on Lötzen.

VIth Corps to advance against the line Lyck-Johannisburg.

The Cavalry of each Army, strongly supported by Infantry, was to cross the frontier on the 13th. The IIIrd Corps of the First Army and the whole of the Second Army were to cross the frontier between the 15th and 17th August. The attack on the Angerapp position North of the Masurian Lakes was not, however, to commence until the IVth Corps (4 days behind the IIIrd Corps) and the XXth Corps had come up into line. The new Army at Warsaw was to throw out its advanced units as if about to attack Austria. He pointed out that the XVIIIth Corps of this Army could not be expected in Warsaw until the 26th August, which was later than Januschkevitch had calculated upon.

This piecemeal offensive was not accepted by Januschkevitch, as being too dangerous. The plan, however, did not envisage an attack. It was in the nature of a threat, and might possibly have produced some effect upon the general situation. In any case, the advance entailed a commencement five days before the rear services of the two Armies were ready.

On the 10th August, Januschkevitch sent out an Instruction ordering an advance of the First and Second Armies.

Januschkevitch's Instruction.

"Germany has directed her main forces against France. Against us there may be assumed to be Four German Corps—Ist, XXth, XVIIth, and Vth, together with some Reserve Divisions and Landwehr Brigades. On our side, by the evening of the 12th day of Mobilisation (11th August), there will be concentrated in the First Army—All the Cavalry (5½ Divisions), IIIrd and IVth Corps, 5th Rifle Brigade, 28th Division (XXth Corps), and possibly Two Regiments of the 29th Division (XXth Corps)—in all, 96 Battalions and 132 Squadrons. The Guard and Ist Corps will not be included in the First Army, and it may be taken that this Army will have completed its concentration by the evening of the twelfth day (11th August).

"The Second Army, except the XIIIth Corps, will complete its concentration earlier than the First Army. This Army will, by the evening of the 12th day (11th August), be composed as follows—Four Cavalry Divisions, the IInd, VIth, XVth, XXIIIrd Corps, and the 1st Rifle Brigade—in all, 136 Battalions and 96 Squadrons. If Two Divisions and the 1st Rifle Brigade are excluded for the defence of Novo Georgievsk, the Second Army will contain 112 Battalions and 96 Squadrons.

"Thus the total of the Armies of the North-West Front, by the evening of the 12th day (11th

August), will be 208 Battalions and 228 Squadrons of field troops to oppose 100 Battalions of German field troops disposed in Four Corps.

"Taking into consideration that war was first declared against us, France as our Ally has considered it her duty to advance against Germany immediately, in order to support us. Naturally, it is our duty to support France in view of the great stroke prepared by Germany against her. This support must take the form of the quickest possible advance against Germany, by attacking those forces left behind in East Prussia.

"The First Army should commence its advance with the object of drawing upon itself the greatest possible enemy strength. The advance must be carried out North of the Masurian Lakes, turning the enemy's left flank.

"The Second Army should advance round the South of the Masurian Lakes, with the object of destroying those German Corps concentrating between the Vistula and the Lakes, and of preventing the Germans from withdrawing to that river.

"The closest liaison must be maintained between the First and Second Armies, a sufficient screen being placed in front of the Lake position.

"The Advance of the Armies of the North-West Front should be in a position to commence on the 14th day of Mobilisation (13th August)."

Januschkevitch's plan was, in fact, to clear East Prussia with the First and Second Armies, prolonging the advance on the left with the new Army, when these two Armies drew up level with Warsaw.

The rear services of the two Armies were not due for completion until the 20th day of Mobilisation (19th August), and so the plan certainly envisaged a start being made without them.

It will be noted that the objective of the Second Army was given as "Destroying those German Corps concentrating between the Vistula and the Lakes, and of preventing the Germans from withdrawing to that river."

This unfortunately suggested to Jilinsky a wider turning movement than that which he had at first proposed. On 12th August he wired that he proposed to extend the left of the Second Army to Chorzele instead of Lomja, in order to have "a greater strategic effect."

Januschkevitch acquiesced.

4. Jilinsky's Instructions to the First and Second Armies.

On the 13th August, from Volkovisk, General Jilinsky issued two Instructions. Both are timed 8.50 p.m. They ran as follows :—

Instruction No. 1 to the Commander of the First Army.

"The enemy has directed his main forces to his Western Front against France. According to our information, he has retained in East Prussia 3 to 4 Corps, with some Reserve Divisions and Landwehr Brigades. His Advanced Guards have moved up to the Frontier, but his Main Force is certainly situated behind the Lakes.

"I intend to assume a determined offensive against him, with the object of cutting him off from Königsberg and seizing his line of retreat to the Vistula.

"For this purpose, the First Army will advance from the line Wierzbolowo–Suwalki to the line Insterburg–Angerburg, turning the Masurian Lakes by the North. The Second Army will advance from the line Augustow–Grajewo–Myszyniec–Chorzele to the line Lötzen–Rudczanny–Ortelsburg and farther to the North, the main strength of the Army being directed on the line Rudczanny–Ortelsburg, in flank and rear of the line of Lakes.

"Thus the Advance will be carried out by turning both flanks of the enemy situated in the Lake area.

"Having at its disposal a large body of Cavalry, the First Army, after crossing the Angerapp, will turn the enemy's left flank by as deep a movement as possible, with the object of cutting him off from Königsberg.

"To the Cavalry I give the task of screening and hiding from the enemy the direction of march of our Corps, the consolidation of specially important points, the seizing of crossings, and the penetrating far in rear of the enemy, in order to prevent the removal of rolling-stock from the railways.

In pursuance of the above order, I direct the First Army to push Cavalry, supported by Infantry detachments, across the Frontier on the morning of the 16th August, driving back the enemy's advanced troops and consolidating the points occupied. On the 17th August all three Corps belonging to the First Army will cross the Frontier, directing their advance firstly on the line Insterburg–Angerburg, and later to turn the enemy's left flank. During this advance a sufficient screen must be maintained towards Lötzen, from whence a German advance is possible.

"The Second Army will advance with its right flank Corps from Augustow on Lyck, Arys, and Lötzen, crossing the Frontier on the 19th August.

"The enemy will be attacked energetically and determinedly whenever and wherever met. This Instruction does not include the fortress of Lötzen. If this place is to be attacked, special orders will be issued.

"The dividing line between the First and Second Armies will be the line Lipowka–Polommen–Soltmahnen–Lötzen inclusive to the Second Army."

Commander-in-Chief of the North-West Front.
General of Cavalry JILINSKY.
Chief of Staff.
Lieut.-General ORANOFSKI.

Instruction No. 2 to the Commander of the Second Army.

"The enemy has directed his main forces to his Western Front against France. According to our information, he has retained in East Prussia 3 to 4 Corps, with some Reserve Divisions and Landwehr Brigades. His Advanced Guards have moved up to the Frontier, but his Main Force is certainly situated behind the Lakes.

"I intend to assume a determined offensive against him, with the object of cutting him off from

Königsberg and seizing his line of retreat to the Vistula.

"For this purpose the First Army will advance from the line Wierzbolowo–Suwalki to the line Insterburg–Angerburg, turning the Masurian Lakes by the North. The Second Army will advance from the line Augustow–Grajewo–Myszyniec–Chorzele to the line Rudczanny–Passenheim, and farther to the North in flank and rear of the Lakes.

"Thus the Advance will be carried out by turning both flanks of the enemy situated in the Lake area, the First Army by the North and the Second Army by the South.

"In pursuance of the above order, I direct the Second Army to push Cavalry, supported where possible by Infantry, across the Frontier along the whole front indicated, on the 18th August. The main bodies of Corps will cross on the 19th August, directing their march to the West of the Masurian Lakes, the main strength of the Army from the line Myszyniec–Chorzele to the line Rudczanny–Passenheim and farther to the line Rastenburg–Seeburg.

"Owing to the particular importance of the Augustow–Grodno area during the impending operations and the unfinished state of the fortress of Grodno, I consider it essential that the IInd Corps should cover this area in its advance Westwards from Augustow. I attach particular importance to the energetic execution of this advance during the first few days of operations, in order to ascertain what forces the enemy has at his disposal.

"For the impending operations, advantage may be taken of the 2nd Division now at Novo Georgievsk, leaving the 1st Rifle Brigade and the Brigade of Frontier Guards as garrison of the fortress. In addition, the 79th Division, due to concentrate at Malkin on the 22nd day of Mobilisation (21st August), will be directed on Novo Georgievsk without detraining at Malkin.

"During the advance to the West of the Masurian Lakes, it is essential to keep in view the possibility of an enemy attack from the direction of Allenstein against the left flank of the Second Army, and therefore the most intensive reconnaissances will be carried out in this direction.

"To the Cavalry I give the task of screening and hiding from the enemy the direction of march of our Corps, the consolidation of specially important points, the seizing of crossings, and the penetrating far in rear of the enemy, in order to prevent the removal of rolling-stock from the railways.

"The First Army will cross the Frontier with its main force on the 17th August, the left flank Corps being directed from Suwalki on Angerburg with protection towards Lötzen.

"The enemy will be attacked energetically and determinedly whenever and wherever met. This Instruction does not include the fortress of Lötzen. If this place is to be attacked, special orders will be issued.

"The dividing line between the First and Second Armies will be the line Lipowka–Polommen–Soltmahnen–Lötzen inclusive to the Second Army."

Commander-in-Chief of the North-West Front.
General of Cavalry JILINSKY.
Chief of Staff.
Lieut.-General ORANOFSKI.

5. Comments upon Jilinsky's Instructions.

The form in which these Instructions are drawn up is open to criticism. Russian Orders and Instructions are always more verbose than our own, but these contain many repetitions, and are confused by not being divided up into paragraphs dealing with each Army in turn. The issue of separate Instructions, in which many of the directions necessary for both Armies are not identical, must lead to confusion. For instance, in Instruction No. 2, Samsonoff is given a final objective of Rastenburg–Seeburg, whereas, in Instruction No. 1, Rennenkampf is told that the Second Army is to move to the line Rudczanny–Ortelsburg. For simplicity and for future reference it is preferable to prepare one Instruction, setting out the general plan which all must know, adding for each subordinate any special matters which he alone requires to know. Remarks upon the ordinary duties of troops, such as Jilinsky issued on the subject of Cavalry action, are unnecessary.

It is interesting to compare the form of Jilinsky's Instructions with those of Von Molkte, issued for the advance of the Prussian Armies into Bohemia in 1866, where the circumstances were very similar. These ran as follows :—

General Order.

> "His Majesty orders both Armies to enter Bohemia and to seek their junction in the direction of Gitschin."

Instruction to the Second Army Commander.

"His Majesty's order to move into Bohemia has been given in to-day's cipher telegram just sent off.

"The direction of Gitschin has been given for the eventual union of the two Armies, because the distances and the communications by road and railway make it seem suitable.

"Of course, this does not mean that Gitschin must be reached in any circumstances. That depends upon circumstances.

"According to all intelligence that we have, it is quite improbable that the Austrian main force could be assembled ready in Northern Bohemia within the next few days. As we are taking the initiative, there might easily be an opportunity of attacking them with superior forces while they are divided, and of following up the victory in another direction. All the same, the union of all our forces for the decisive battle must all the time be kept in view. The Commander of each Army, from the moment when he is facing the enemy, must employ the troops entrusted to him according to his own judgment and the requirements of the situation, but in so doing must always consider the situation of his fellow Army. Mutual support will be facilitated by their continuously comparing notes."

To the copy forwarded to the First Army Commander were added these words:—

"As the weaker Second Army has the more difficult task of issuing from the mountains, it is the more incumbent on the First Army, as soon as ever its junction with the force of General Herwarth has been effected, to shorten the crisis by its swift advance."

6. Strengths of the Opposing Forces as they entered the Campaign.

(i) *Russians.*

First Army—IIIrd, IVth, and XXth Corps.
5th Rifle Brigade.
1st and 2nd Guard Cavalry Divisions.
1st, 2nd, and 3rd Cavalry Divisions.
1st Independent Cavalry Brigade.

In all—86 Battalions of First-Line Infantry.
120 Squadrons of Cavalry.

Second Army—Ist, IInd, VIth, XIIIth, XVth, and XXIIIrd Corps.
1st Rifle Brigade.
4th, 6th, and 15th Cavalry Divisions.

In all—164 Battalions of First-Line Infantry.
110 Squadrons of Cavalry.

The total of the forces of the North-West Front, destined for immediate operations in East Prussia, was thus :—

250 Battalions of First-Line Infantry.
230 Squadrons of Cavalry.

It will be seen that the Russians were making no use of their Second-Line troops in the initial advance. This was due to the fact that they were not ready. But Corps were also very short of their full establishments. Had they been up to strength, the 9½ Corps of the two Armies would have produced 304 Battalions instead of 250. The Armies therefore started with a shortage of just under 18 per cent of their full complement of Infantry. This was due to the fact that First-Line Battalions were required for duty in rear until they could be relieved by Second-Line units. The Russians were advancing from what are now Lithuania and Poland, where there were many persons ill-disposed to the Russian Government. The guarding of depots, detraining stations, bridges, and railway lines had to be carried out at the moment by First-Line troops. Local detachments, such as were used by the Germans, were not available.

(ii) *Germans.*

Ist, XVIIth, XXth, and Ist Reserve Corps.
Fortress Garrisons.
1st Cavalry Division.

In all—144 Battalions of Infantry (96 First-Line).
84 Squadrons of Cavalry.

Comparing these figures with Januschkevitch's calculations, we find that the Russians produced a superiority of 42 Battalions, and the Germans one of 44 Battalions, over his figures.

The actual superiority in favour of the Russians, therefore, works out as 1·7 to 1 instead of 2·7 to 1 as regards Infantry, and 2·7 to 1 instead of 8·5 to 1 as regards Cavalry.

The initial Russian contingent must have fallen short of the agreed upon 800,000 men by at least 50 per cent.

7. A suggested Instruction to be issued by Jilinsky.

VOLKOVISK, *13th August.*

1. The enemy has directed his main forces to his Western Front against France, and has retained in East Prussia three or four Corps, some Reserve Divisions and Landwehr Brigades. German advanced troops have moved up to the frontier, but the enemy's main force is situated to the West of the Masurian Lakes.

2. The First and Second Armies will advance into East Prussia with the object of cutting off the Germans from the Vistula.

The advance will be carried out as follows :—

(i) *First Army*—IIIrd, IVth, and XXth Corps.
5th Rifle Brigade.
1st and 2nd Guard Cavalry Divisions.
1st, 2nd, and 3rd Cavalry Divisions.
1st Independent Cavalry Brigade.

Westwards towards the Insterburg Gap.

Objective: To draw the German forces as far Eastwards as possible, with a view to allowing the Second Army to operate against their rear. While operating with vigour, the First Army will not allow itself to be drawn into a decisive battle to the East of the Masurian Lakes. It is important to sever the enemy's communications with Königsberg at an early date.

Main bodies of Corps will cross the frontier on the 21st August.

(ii) *Second Army*—I[st], VIth, XIIIth, XVth, and XXIIIrd Corps.
1st Rifle Brigade.
4th, 6th, and 15th Cavalry Divisions.

Northwards from the line Myszyniec–Chorzele to the line Rastenburg–Seeburg.

Objective: To attack the rear of the German forces opposing the First Army and cut them from the Vistula. The initial stages of the advance will be rapid. The left flank will be protected against possible enemy attacks from the direction of D. Eylau.

Main bodies of Corps will cross the frontier on the 23rd August.

(iii) *IInd Corps.*

Will remain under the orders of the N.W. Front.

Objective: To drive the enemy to the West of the line of the Lakes, and to occupy the Gaps from Lötzen to Johannisburg inclusive.

Orders will be issued later for the reduction of Fort Boyen.

Main body of Corps will cross the frontier on the 21st August.

3. The Second Army will be responsible for the maintenance of an adequate force for the defence of the fortress of Novo Georgievsk. The 79th Division, due to arrive at Malkin on the 21st August, will be directed on Novo Georgievsk without detraining at Malkin.

4. Headquarters N.W. Front will be established at Volkovisk from to-day.

CHAPTER IV.

Discussion of the Russian Plan.

Map I.—Strategical Map.
Map II.—General Map of the Advance of the Russian Armies.

1. German Situation.

THE Strategy of a simultaneous war in both East and West had long been debated in Germany. This Strategy had been embodied in a plan prepared by Graf von Schlieffen, the Chief of the General Staff.

In 1901 an important War Game was played under the supervision of Von Schlieffen. It was desired to gain more time in the East than could be obtained by a mere defensive attitude of the German forces left in this area. The project put forward in the War Game was an offensive with Seven Corps against Grodno, with the object of upsetting the Russian concentration. Could the Petrograd–Vilna–Grodno and the Baranovitchi–Bialystok railway lines be thoroughly interrupted, it was thought that the Russian Army in the Vilna area could be separated from the Russian forces in Poland. Having upset the Russian concentration, the majority of the Seven German Corps were then to be transferred to the French front to take part in the decision in the West. Such a project possessed great possibilities, but the increasing preparedness of France rendered the project more dangerous each year. The Russians were aware of this project, and their nervousness with regard to the direction of Grodno is seen in all Instructions issued by the Russian Commanders. By 1905 the German plan consisted in employing the mass of their forces in the West, while leaving Thirteen Infantry and Two Cavalry Divisions to oppose Russia. The decision to attack France first was arrived at as the only means of obtaining an immediate success, and so of counteracting the superior but more slowly developing man-power of Russia. To attack Russia first might lead to a traditional withdrawal on the part of the Russians, and a lengthy campaign in consequence. By 1914 this Eastern contingent had been reduced by Von Molkte, who had succeeded Von Schlieffen, to Eleven Infantry Divisions and One Cavalry Division. French Mobilisation had been increasing in efficiency, and it was important to be absolutely sure of a quick decision in the West. Von Moltke hit upon the idea of employing his Reserve Formations in almost the same manner as his First-Line Troops, and he thus increased the size of his initial contingents on both fronts, surprising his opponents in this respect. As regards the Eastern contingent, there were to be Six First-Line and Five Second-Line Divisions. Von Molkte proposed to run greater risks in the East, for his contingent was weaker in both quantity and quality to that proposed by Von Schlieffen, and the Russians had been improving their Mobilisation arrangements, though not to the same degree as the French.

The German Commander in the East, General Von Prittwitz, had a difficult situation to meet.

His objective was to delay the Russians until he could be supported from the West. His delaying operations were to be of no short duration, for it would be weeks before a decision could be obtained in the West. He was greatly inferior in numbers to the Russian forces which might advance against him, and the territory he had to defend was geographically most unsuitable. It was desirable, however, to prevent a Russian invasion of East Prussia if possible.

The enemy was known to be more prepared in the Vilna area than elsewhere, and the Eastern portion of East Prussia was the most likely to be invaded first. To meet such a Russian advance a German concentration far to the East was necessary, with a consequent exposure of its right flank and rear to an advance from Poland. Von Prittwitz had located his forces in two groups, observing the two directions from which he might be attacked, while maintaining the remainder, a little less than half his whole force, in a central position ready to reinforce either group as required. His troops were superior in training and equipment, though inferior in numbers. He had a great advantage in mobility, by reason of his motor transport and the excellent road and railway system existing in the German portion of the theatre of operations.

2. Russian Situation.

The Russians were making an eleventh-hour change of plan, with all its consequent difficulties. They were about to improvise a second offensive campaign on a large scale. All the conditions of the moment were against them, with the sole exception of numbers. They were already committed to an offensive against Austria, and resources in technical material, artillery, transport, and signal stores were insufficient for a second extensive offensive. They were being urged to hurry, while the concentration of their troops could not be hastened by any mechanical means in their power. They were further being asked to advance in a direction unsuited to the actual concentration under way.

Januschkevitch, in his wire to Jilinsky, had said that an advance was only to take place when sufficient troops had been collected. It is true that the troops would be in their concentration areas in more or less complete condition about the 13th August, but the services which were to maintain them in the field would not be ready until the 19th August. The Russian Supreme Headquarters was deliberately starting off the two Armies upon an advance, which was to continue for weeks, without their proper means of supply.

Januschkevitch's plan consisted in pushing forward the First Army as a bait to draw the German forces eastwards, and in then falling upon their rear with the Second Army. He compromised to the desire of the French for an advance against Posen by the formation of a new Army in Warsaw.

The operation of the concentric advance of two separate forces is a classic one, of which there are many examples in History, notably the Prussian Campaign against Austria in 1866. All writers on Strategy point out the difficulty of manœuvring separated forces against a concentrated enemy, but the advent of modern inventions—motor transport, aeroplanes, wireless, and telephones—has considerably lightened the task of a commander in such an operation. Unfortunately, it was just in these very facilities that the Russians were deficient, and conditions with them more nearly approximated to those of older days. On the other hand, the Germans were equipped as a modern Army.

The essence of the strategy of a concentric advance is careful timing. If one Army is threatened by superior force, it can be directed to advance warily, refusing battle, and even withdrawing if necessary, to continue the advance at increased speed when the danger has passed, or when the other Army is similarly threatened. Only thus can the divided forces be brought eventually into tactical supporting distance on the battlefield.

On the other hand, the inferior Army must either take its courage in its hand and attack the concentrically advancing forces of the enemy in detail, while there is yet elbow-room to manœuvre, or withdraw to escape the jaws of the pincers descending upon it. The cardinal rule is that the more menacing force must be dealt with first, the other being attacked later.

Taking the configuration of the Frontier, the original plan of concentration, and the need for haste, such a concentric advance was imposed upon the Russians. The Russian railways were not situated so that any lengthy train movements from one flank to another could be carried out.

If efficiently executed, the plan offered the chance

of a crushing success by the envelopment of the inferior enemy.

The situation of the Russians at the commencement of the operation was as follows :—

First Russian Army.

The distance from the Frontier to the objective line Insterburg–Angerburg is 37 miles. Two railways run up to the Frontier on either flank of the Army front. That at Wierzbolowo reaches the Frontier, and that at Suwalki reaches within 15 miles.

The Roads—

(*a*) Mariampol–Kalwarja–Suwalki,
(*b*) Mariampol–Wylkowyzki–Wierzbolowo,
(*c*) Kalwarja–Wysztyniec,

were in good condition.

The front from Wierzbolowo to Suwalki is about 37 miles, narrowing on the objective line to about 31 miles.

The First Russian Army was formed of Three Corps.

The XXth Corps was stationed in peace time in the area of Riga. It formed part of the Vilna Military District. It had been due to concentrate at Lyublin in the Fourth Army on the Austrian front. It was concentrated at Kovno. The railway journey entailed was 280 miles, being along the main line. The distance by the main road from Kovno to the Frontier at Eydtkuhnen is 56 miles.

The IIIrd Corps was stationed in peace time in the area of Vilna. It formed part of the Vilna Military District. It was due to concentrate at Orany, Olita, and Simno, all situated on the railway line to Suwalki. The railway journey entailed was 50 miles. By road, from the head of the concentration, Simno, to the Frontier at Wisztyniec is 45 miles. The roads are passable and well-kept.

The IVth Corps was stationed in peace time in the area of Minsk. It formed part of the Vilna Military District. It was due to concentrate at Druskienki on the R. Niemen, lying 12 miles by good road from the main Grodno-Vilna railway. The railway journey entailed was 180 miles. By road, *viâ* Suwalki, to the Frontier at Filipowo is 65 miles. The roads are passable and well-kept.

The Cavalry was maintained at practically war strength, close to the Frontier, and could be quickly concentrated.

The concentration of the First Russian Army offered no great difficulty, and was being carried out according to long prearranged plan. Roads and railways admitted of the rear services being moved up behind the units they were to serve.

The Three Corps all belonged in peace time to the Vilna Military District, of which Rennenkampf was the Commander.

Second Russian Army.

To get from one flank of the Army to the other, the only railway line available runs through Novo Georgievsk, Bialystok to Mlawa being 200 miles. All the roads in the Russian area north of Ostrolenka are sandy and unmetalled.

The front of the main advance is 31 miles, remaining about the same on the objective line.

The distance from the Frontier to the objective

line Rastenburg-Seeburg is 43 miles. The only railway, that to Ostrolenka, lies 31 miles from the Frontier. The lines which reach the Frontier at Grajewo and Mlawa lay outside the Army frontage, and were of little use for the concentration or supply of the Army.

The Second Russian Army was composed of Five Corps.

The IInd Corps was stationed in peace time in the area of Grodno. It formed part of the Vilna Military District. General Scheidemann had been in command of the Corps since 1912. The Divisional Commanders, Paretsky and Slusarenko, had been in command of the 26th and 43rd Divisions since 1911. The Corps was due to concentrate at Augustow, which lies on the loop-line to Suwalki. The railway journey entailed was 40 miles. By road from Augustow to the Frontier opposite Lyck is 12 miles.

The VIth Corps was stationed in peace time in the area of Bialystok. It formed part of the Warsaw Military District. General Blagovestjensky had been in command since 1912. The Divisional Commanders, Komaroff and Richter, had been in command of the 4th and 16th Divisions since 1912 and 1910 respectively. The Corps was due to concentrate along the R. Bobr. Its right could be served by the Bialystok-Grajewo railway, while its left lay near Lomja. The concentration was to be carried out by road. From Osowiec, the centre of the line, the distance by road to the Frontier is 17 miles.

The XIIIth Corps was stationed in peace time in the area of Smolensk. It formed part of the Moscow

Military District. General Klueff only assumed command of the Corps on its arrival at Bialystok. The Divisional Commanders, Ygrumoff and Prejentseff, had been in command of the 1st and 36th Divisions since 1910 and 1913 respectively. It was due to concentrate at Bialystok, entailing a railway journey of 300 miles. By road, the distance to the Frontier at Bialla is 60 miles.

The XVth Corps was stationed in peace time in the area of Warsaw. It formed part of the Warsaw Military District. General Martos had been in command since 1911. The Divisional Commanders, Torklus and Baron Von Fittinghof, had been in command of the 6th and 8th Divisions since 1910 and 1913 respectively. It was due to concentrate at Zambrowo, about 13 miles by good road from the main Bialystok-Warsaw railway. The railway journey entailed was 80 miles. From Zambrowo to the Frontier near Kolno is 37 miles.

The XXIIIrd Corps was stationed in peace time in the area of Warsaw. It formed part of the Warsaw Military District. General Kondratovitch had only just taken over command. The Divisional Commanders, Mingin and Sirelius, had been in command of the 2nd and 3rd Guard Division since 1913 and 1910 respectively. The Corps had been broken up on Mobilisation. The Corps Cavalry went to the Austrian Front, the Howitzer Division to the IInd Corps, and the Sapper Battalion to Grodno. All the rear services were distributed amongst various Corps on the Austrian Front. It was due to concentrate at Bialystok and Sokolka, with units at Volkovisk. The railway journey entailed was 120 miles.

The Ist Corps was stationed in peace time at St Petersburg. It formed part of the St Petersburg Military District. General Artamonoff had been in command since 1911. The Divisional Commanders, Dushkevitch and Reshikoff, had been in command of the 22nd and 24th Divisions since 1912 and 1910 respectively. It had been due to concentrate at Kovno to form part of the First Army, but was diverted to the Ninth Army and ordered to concentrate at Warsaw. It was then drawn into the Second Army by Samsonoff, with the unwilling consent of Jilinsky. The railway journey entailed was 730 miles.

Including the Ist Corps, it should be noted that Samsonoff's Army was made up of Corps from four Military Districts, and therefore not so homogeneous as that of Rennenkampf.

The Corps of the First Army were concentrating along the Frontier facing the Insterburg Gap, and at distances varying from four to five days' march from the Frontier.

The Corps of the Second Army were concentrating opposite the Southern half of the line of the Masurian Lakes. The concentration favoured a move between Lötzen and Johannisburg. Four Corps—the IInd, VIth, XIIIth, and XVth—were in front line at distances of one to five days' march from the frontier, with the mutilated XXIIIrd Corps slightly behind and in the centre.

The first plan proposed by Jilinsky, although it contemplated an advance before the rear services were ready, was only in the nature of a threat, and did not involve a continuous advance at that early date. It might conceivably have had some effect upon the general situation. A detailed plan for the operations of the IInd and VIth Corps direct against the Southern end of the Lakes must have existed, for such an advance formed part of Case G. But Jilinsky rejected any such plan, and commenced to counter-march the Second Army several days' march to the West, with a view to gaining more strategic effect, and drawing clear of the difficult Lake country.

As has been pointed out, the timing of a concentric movement is difficult. Januschkevitch had left the timing of the operation to Jilinsky, merely indicating to him that the troops should be ready in their concentration areas on the 13th August. One would have expected Jilinsky, after consultation with his Army Commanders, to point out to Supreme Headquarters that the troops would not be ready until the 19th, unless it were possible to hasten the movements of the rear services in some way. Jilinsky evidently accepted the fact that an immediate advance was of the utmost importance, but by counter-marching the Second Army to the West he made it still more impossible for the rear services of that Army to function.

Jilinsky ordered the First Army to cross the Frontier on the 17th August, and the Second Army on the 19th.

If we assume that Jilinsky's timing were carried out, the following situations might have arisen :—

(i) The Germans might decide to oppose Rennenkampf first while holding off any Russian advance from Poland.

Rennenkampf's orders were to push straight on as fast as he could, driving back the enemy wherever found. From the frontier to the R. Angerapp is 24 miles. Calculating on marches of 12 miles a day, a battle on the R. Angerapp might be expected from the 19th August onwards.

Samsonoff might reach his objective, Rastenburg–Seeburg, in from three and a half to four days. Taking into account that the Second Russian Army would have been marching for over a week since detrainment, that the roads were very bad, and that even weak German forces, operating against the Russian left flank, would slow up the general movement, in all probability Samsonoff would not reach his objective until the 23rd or 24th August. At this moment he would have cut the German communications to the Vistula through Allenstein, but there still remained the line through Königsberg–Braunsberg–Elbing, lying a farther 50 miles to the North. His right would lie 18 miles from Angerburg, the point at which Rennenkampf might hope to emerge from the Insterburg Gap into the area of operations of the Second Army. Relying on their Northern line, therefore, the Germans had five days at least in which to deal with Rennenkampf before being forced to turn and face Samsonoff.

(ii) The Germans might decide to oppose Samsonoff first, while holding off Rennenkampf.

If the Germans decided to concentrate South of the Allenstein railway, in the region of Passenheim–Hohenstein, they might expect to meet Samsonoff on the 21st August. At this moment Rennenkampf's left might be emerging from the Insterburg Gap,

45 miles away from Passenheim. In this case the Germans would have no fear as regards their communications to the Vistula.

From these calculations the difficulties inherent in such an advance as the Russians were contemplating become apparent. The enticing of the Germans Eastwards was the basis of the Russian plan to cut them off from the Vistula; an early advance on the part of Rennenkampf was therefore essential. The general timing of the operation, however, depended upon the dates to which the Second Army could work, for it was with this Army that the decisive blow was to be delivered. If the Second Army were to be delayed, the situation of the First Army, with orders to push straight ahead and attack the enemy wherever found, might become exceedingly critical. As will be seen later, Jilinsky's timing of his counter-march to the West for the Second Army was too optimistic, and the Army could not keep up to the time-table. The interval of time between the dates on which the two Armies were to cross the Frontier, therefore, became greater than the forty-eight hours ordered by Jilinsky. The timing of the operation was upset from the outset, and the Second Army was continually being urged to keep up to an impossible table of marches.

The demands of Russia's Allies in the West were insistent upon the question of an early advance on her part, but nothing could justify a start being made with such deficiencies that there was danger of a breakdown. What actual delay was necessary is difficult to calculate without a knowledge of all the circumstances. It should be remembered that the rear services in general were not due for completion until the 19th August. As the Second Army had to carry out its counter-march by road, there might have been some chance of altering the destinations of the rear services to suit the new areas, and so saving a little time. A delay of six days would have ensured everything being ready, but this delay might possibly have been reduced to one of four days.

3. Comments upon the Russian Advance.

The Russians were about to carry out an advance comparable in size to that of the German right wing in France, directed over very difficult country with inadequate rear services. The final order for the advance was only issued while the troops were marching to their concentration areas, after considerable change in the order of battle. The German advance had been thought out and perfected in every detail for years before the date chosen for its commencement, and was to be directed by a Staff incomparably superior to the Russian Staff, while being supported by every means of transport existing.

As has already been pointed out, the French plan for an early offensive in Lorraine was based upon the retention of a sufficient German force in the East. In order to effect this, a Russian offensive against Germany had been agreed upon, to commence as nearly simultaneously as possible with the French advance. The Russians had made no preparations for such an advance, and this fact had evidently become known in Berlin, for the Germans reduced their forces in the East below the minimum calculated upon by the French.

The Russians had failed to fulfil their engagements for two main reasons.

The first was one of time. In actual practice there can be no quick change in a programme of concentration. Railway time-tables for such a movement as that of a General Mobilisation require checking and rechecking from many points of view. Every order for every train must lie ready filed in the office of the Command concerned, and one general telegram sets the whole system rolling. Improvisation on such a vast scale is impossible, more especially where great distances are involved, as was the case in Russia, and with a great deficiency of rolling-stock on the railways.

The second is one of psychology. In every nation General Mobilisation is somewhat of a gamble. There is always an anxiety as to whether it will be a success. If Mobilisation does not run smoothly according to prearranged plan, there may be danger of delays in the concentration taking place later. The population must have its heart in the war to some extent. In Russia the reason for going to war, as understood by the people, was the attempt on the part of Austria to dragoon the Southern Slavs in Serbia. Here was a reason for war which every Slav understood, and the most was made of this national cry in Russia. Austria was the national enemy, and the people expected the main effort to be made against her. It would have been a bitter disappointment to all Russians to find Austria invading Russian territory. From a purely Russian point of view, therefore, it was preferable to make the main effort against Austria.

Lack of preparation and national feeling, therefore, prevented the promises of the Russian General Staff being fulfilled, and the concentration of 1910, according to Case A., was allowed to continue. The Russian General Staff then attempted to keep up to their promises by improvising a fresh campaign against Germany.

CHAPTER V.

The Advance of the First and Second Armies.

Map II.—General Map of the Advance of the Russian Armies.
Sketches A, B, C, and D.

FIRST ARMY OPERATIONS.

1. RENNENKAMPF'S ORDERS.

FROM the early days of August both sides carried out raids across the frontier. The Russian Cavalry advanced considerable distances into German territory, but always retired again. They established the fact that both Divisions of the 1st German Corps and the 1st German Cavalry Division were close up to the frontier, but they never succeeded in penetrating the German screen. This was the state of the information about the enemy when Rennenkampf issued his orders for the advance of the First Army.

These orders ran as follows :—

First Army Order No. 2.

VILNA, 15*th August* 1914.

1. Information has been received that units of both Divisions of the German 1st Corps are concentrating on the line Pillkallen–Pillupönen, with a

Cavalry Division (6 Regiments) on the left flank at Pillkallen.

2. In accordance with Instruction No. 1 of the N.W. Front, the Army will cross the Frontier on the 17th August, and will occupy on that day, with the main bodies of Corps, the line Willuhnen–Stallupönen–Germingkehmen–Dubeningken–Kowahlen.

3. Cavalry—

(*a*) General Oranofski will protect the right flank of the Army, reconnoitring to Tilsit–Kraupischken, occupying Schillehnen by the evening of the 17th August.

(*b*) General Khan Nachivanski and General Gourko will carry out the orders in my Instruction No. 1.

4. The zones of advance, requisitioning, and operations of the Corps will be bounded as follows :—

XXth Corps : *North*—Schirwindt–Willuhnen–Pillkallen–Kussen.
South — Eydtkuhnen–Stallupönen–Gumbinnen, exclusive.

IIIrd Corps : *North*—The Southern line of the XXth Corps inclusive.
South — Kopsodzie – Wicknaweitschen–Enzuhnen–Walterkehmen, exclusive.

IVth Corps : *North*—The Southern line of the IIIrd Corps inclusive.
South — Przerosl–Goldap–Darkehmen exclusive.

5. The 5th Rifle Brigade is placed under the orders of General Gourko.

6. Headquarters of Corps will pass the night as follows :—

XXth—Wladyslawow ; IIIrd—Eydtkuhnen ; IVth —Przerosl.

7. Reports will be sent as follows :—

Up to 3 A.M. on the 17th to Vilna, after that to Wierzbolowo.

8. Communications *viâ* Imperial Telegraph.

9. Next for Command, Lieut.-General Mileant; General of Infantry, Smirnoff.

Instruction No. 1.

To carry out the orders of the Commander-in-Chief, North-West Front, the Commander-in-Chief of the First Army issues the following orders :—

1. The Army Cavalry will advance in 2 groups on the 16th August.

(*a*) The right group—on Insterburg, turning Stallupönen and Gumbinnen to the North.

(*b*) The left group—to protect the left flank of the Army firstly from the direction of Marggrabowa, and secondly from Lötzen.

(*c*) The Independent Cavalry Brigade will protect the right flank of the Army, reaching Schillehnen on the 17th August.

The Cavalry must, moreover, screen the movements of Corps as much as possible, and prevent the removal of rolling-stock from the railways.

2. On the 17th August the main bodies of Corps will reach the line Willuhnen–Stallupönen–Germingkehmen–Dubeningken–Kowahlen.

MILEANT, *Chief of Staff.*

2. Comments on Rennenkampf's Orders.

The six Divisions of the Army were advancing on a front of 35 miles, each Corps being given a lane down which to advance, without reference to the possible positions of the two German Divisions and the Cavalry Division known to be near the frontier. The average strength of the Russian Divisions was 13 Battalions, and they may therefore be taken as the equivalent of German Divisions. At the battle of Mons von Kluck struck the four British Divisions with six Divisions on a frontage of 17 miles, and at Le Cateau he employed the same frontage for eight Divisions. It is considered, therefore, that Rennenkampf's frontage was too great under the circumstances.

The Forest of Rominten lay in the Southern portion of the area over which the Army was advancing. It is very dense and unsuitable for the movement of troops, with only one road running down its centre. Rennenkampf ordered two Corps to march North of the Forest, and gave his third Corps a broad frontage including it. The Commander of the Southern Corps elected to send one Division to the North and the other to the South of the Forest, with one Regiment from his Northern Division down the centre of it. It would have been preferable to have disposed all three Corps North of the Forest, while ordering the Southern Corps to send a Regiment through it. There would still have remained Gourko's detachment marching to the South of the Forest.

By this means the frontage of the three Corps would have been considerably reduced, and Rennenkampf would have been able to concentrate greater strength towards his right flank, where he required it, for his orders were to cut the Germans from Königsberg and to drive them away from their Northern Line of Communications.

There is no finesse in Rennenkampf's plan. He takes no account of the enemy's forces, and does not group his troops, so that they may have the best strategical effect if his manœuvre is successful.

It should be noted that Corps are ordered to locate their Headquarters at definite points. Owing to the lack of signal communications, it was necessary to make use of the Imperial telegraph lines.

Instruction No. 1 to the Cavalry is inadequate, and is nothing short of a delegation of authority to the subordinate Cavalry Commanders. It was not sent to Corps, consequently the IVth Corps did not know what Gourko's detachment was doing. Rennenkampf was apparently relying upon being able to send daily orders to his Cavalry—a method of control which can never lead to the best results. As in the case of all bodies of troops likely to be detached for long periods, Cavalry require the most carefully worded Instructions, informing them what the general plan is, and what is expected of them as regards co-operation. The execution of the task is left in the hands of the detached Commander.

The grouping of the Cavalry is open to criticism. There are three detachments acting directly under the Army. With the three Corps the Army had thus to issue orders to six units, disposed over a wide frontage. Two of the Cavalry detachments were given the definite task of protecting the flanks of the Army, and it would have been preferable to place them under the command of the Commanders of the flank Corps. The flank Corps would then have had three units each, and the Army four units, to which to issue orders. With such a large force of Cavalry at his disposal, and on account of his orders to watch the direction of Lötzen, Rennenkampf's employment of so much Cavalry for protective duties on the flanks may perhaps be accepted. There remained the Cavalry Corps of 72 Squadrons with 48 guns.

Rennenkampf gave the Cavalry Corps no specific Instructions. It was merely told to march on Insterburg, turning Stallupönen and Gumbinnen by the North. It was not even told the hour at which the march of Corps would begin.

The main task of the Cavalry, in the complete absence of aerial reconnaissance, was the collection of accurate information as to the enemy's dispositions. The direction of the march of Corps could then have been arranged so that they could support one another mutually, outflanking the enemy's groups, and forcing them to withdraw or be enveloped. All other tasks were for the moment secondary. Rennenkampf's objective was the line Insterburg–Angerburg. Between these two places ran the R. Angerapp, which was an historic defensive position, the Southern flank of which rested upon an impassable obstacle. The Germans were known to have reconnoitred and prepared this position, and it was essential to ascertain whether they intended to oppose the Russian advance in this position or not, and where their Northern flank rested. The distance from the frontier at which it might be expected to meet strong German forces not being great, and in view of the wide front upon which the Army was moving, it is considered that the Cavalry Corps might have been divided into two Groups, one Group being maintained strong enough to deal with the German Cavalry Division known to be near Pillkallen.

3. Suggested Instruction to the Cavalry Corps.

Instruction to Cavalry Corps.

1. Main Bodies of Corps will cross the frontier at 12 noon on the 17th August.

   The Cavalry Corps will cross the frontier at dawn on the 16th August, disposed as follows :—

Right Group—

Three Cavalry Divisions. Khan Nachivanski.

Objective : Pillkallen. To drive the German Cavalry Southwards.

Information required—

(i) Do the enemy intend to hold the line of the R. Angerapp, or to oppose the Advance to the East of it ?

(ii) Where does his Northern flank rest ?

Left Group—

One Cavalry Division. General Rauch.

Objective : Goldap.

Information required—

Do the enemy intend to hold the line of the R. Angerapp, or to oppose the Advance to the East of it?

2. If the R. Angerapp is found unoccupied, the Right Group will hold the crossings between Insterburg and the point where the Gumbinnen–Insterburg railway crosses the river, and the Left Group the crossings at Darkehmen, until the arrival of the Infantry. Reconnaissances will be sent forward to ascertain whether the Germans are retreating Westwards or are coming up to hold the line of the R. Angerapp.
3. The main line to Insterburg will be cut in succession West of Stallupönen and Gumbinnen, the line from Gumbinnen to Darkehmen, and the lines running out of Goldap, in order to prevent the removal of rolling-stock.
4. Reports will be forwarded through the nearest Infantry Advanced Guards.

4. The Events of 17th August. (Sketch A.)

On the 17th the hours of starting of the Russian Corps differed considerably. Nothing had been laid down on this subject by the Army, and no liaison had taken place between neighbouring Divisions. On the right the XXth Corps crossed the frontier between 12 noon and 1 P.M. The IIIrd Corps, in the centre, crossed between 8 and 9 A.M. In the South, in the IVth Corps, the 40th Division crossed after 11 A.M., and the 30th Division between 1 and 2 P.M.

The IIIrd Corps was the first to come into touch with the enemy near Stallupönen. The 27th Division, on the left flank of this Corps, was ordered to close in to the North and envelop the German right flank. The delay in the start of the 40th Division, and the movement of the 27th Division to the North, caused a large gap in the Russian line between the IIIrd and IVth Corps. Neither Division appears to have taken proper steps to observe this gap. The attack at Stallupönen seemed to be going well until about 4 P.M., when the 27th Division found itself attacked by a strong German force coming up almost directly in its rear.

General Von Falck, who was at Tollmingkehmen, hearing the battle at Stallupönen, sent off at 11.30 A.M. 4 Battalions and 5 Batteries to attack the Russians advancing against Stallupönen. He left 2 Battalions, 4 Squadrons, and 1 Battery to oppose the Russians on his immediate front, with orders to retire slowly on Tollmingkehmen if seriously attacked.

Von Falck's attack came in towards Göritten directly in rear of the Russians. The effect was immediate. In the 105th Orenburg Regiment the Russians laid down their arms almost by companies; over 3,000 prisoners were taken, and the 27th Division withdrew once more to the Frontier in considerable disorder.

During the night of the 17th/18th the Germans effected their withdrawal towards Gumbinnen. They had disorganised the centre of the Russian Advance, inflicted severe casualties upon them, and taken a large number of prisoners. Their casualties were insignificant, though they lost 7 field-guns and some waggons.

The remainder of the Russian advance encountered little opposition, the objectives for the day being duly reached.

The Russian Cavalry Corps, under Khan Nachivanski, gained no contact with the enemy, moving wide to the North of Pillkallen. The Khan reported pathetically, "The enemy avoided mounted action, and retreated quickly." The Cavalry Corps covered some 14 miles on this day.

5. German Dispositions. (Sketch A.)

The German dispositions for this Frontier Protection may be taken as a good example of what this kind of defence should be.

All roads leading to the Frontier were watched by motor detachments, provided with quick means of communication, to the Headquarters of the various Frontier Protection Detachments.

These were situated as follows :—

(*a*) Goldap . .	5 Battalions Infantry. 2 Batteries Artillery.
With advanced posts at—	
Kowahlen .	1 Machine-Gun Company. 3 Batteries Artillery. 2 Squadrons Cavalry.
Gurnen . .	1 Battalion Infantry. 1 Battery Artillery. 1 Squadron Cavalry.

(*b*)	Tollmingkehmen	6 Battalions Infantry. 6 Batteries Artillery. 1 Squadron Cavalry. 1 Pioneer Company.
(*c*)	Stallupönen .	9 Battalions Infantry. 12 Batteries Artillery. 4 Squadrons Cavalry. 1 Company Pioneers.
(*d*)	Pillkallen .	1 Cavalry Division.
(*e*)	Tilsit . .	1 Landwehr Brigade.

Thus a front of 42 miles was observed by a Corps and a Cavalry Division, the Cavalry being placed in the North on the exposed Flank. Detachments were of all arms, and sufficiently strong to prevent their being brushed aside before they could obtain the necessary information.

The German Commander, General Von François, appears to have had a definite programme of withdrawal. Detachment Commanders were permitted to undertake local offensives, within the limits of the time allowed, should the opportunity appear favourable. The German troops were handled with great coolness, Von François keeping to his programme despite the peremptory orders from his Chief, General Von Prittwitz, to break off the fight and withdraw to Gumbinnen.

6. The Events of 18th August. (Sketch A.)

No attempt was made by the Russians on the 17th to follow up the retreating enemy, and the effect produced by the disaster in the 27th Division delayed the sending out of orders for the 18th until the early hours of that day. A start was therefore not made until late.

On the night of the 18th the advanced troops reached the line—

Mallwischken – Mingstimmen – Stallupönen – Pillupönen–Rakowken–Seesken.

Only the Cavalry Corps encountered opposition, having skirmishes with the German Infantry and Cyclists at Wittgirren and Mallwischken. The Cavalry Corps completed some 14 miles on this day, finally bivouacking 3 miles North of Mallwischken.

Oranofski's Independent Cavalry Brigade received no orders, and spent the night at Schillehnen.

Orders for the 19th and 20th reached the Russian Corps during the early hours of the 19th, and only in mutilated form. The Corps receiving stations must have been at fault, for the Germans picked up these wireless orders correctly, and were thus fully informed of what was about to occur.

The line to be occupied on the 19th was given as—

Uszballen – Karmohnen – Puspern – Sodehnen–Goldap.

A general halt was to take place on the 20th.

All the Corps Commanders were now reporting that their rear services were not working well. Communications were blocked with undelivered telegrams, and there was a general state of disorder. The German inhabitants were flying in front of the Advance, driving their flocks and herds in front of them, and progress was therefore becoming very slow. The effect of having started with incomplete rear services was making itself acutely felt after only a few days' marching. Rennenkampf therefore ordered a general halt for the 20th August.

7. The Events of 19th August. (Sketches B. & C.)

On the morning of the 19th the XXth Corps (28th and 29th Divisions) and the IIIrd Corps (25th and 27th Divisions) moved forward to take up the line Uszballen–Puspern–Sodehnen.

The Cavalry Corps, hearing that Kraupischken was occupied by German Infantry, moved forward to that place, coming into contact with Landwehr units (2nd Landwehr Brigade) at Kauschen. The Landwehr were driven back across the R. Inster, losing two guns and some prisoners. They must have suffered heavily in this action, as evidenced by the German graves near the roadside at Kauschen. At darkness the Cavalry Corps withdrew far to the North to Lindental, taking no steps to cover any of the roads towards Uszballen, the right flank of the Army.

This affair was a brilliant little Cavalry action, though carried out against troops of second category. It is interesting to note that the two guns were captured by Count Wrangel, then commanding a Squadron of Guard Cavalry, and later to become Denikin's successor in South Russia.

As a general rule, Cavalry withdraw at night from the zone between two Armies about to become closely engaged, in order that they may not be at the mercy of an enemy night attack. They withdraw either to the flank of or behind the troops they have been covering during the day. The operation must be carefully timed, and full information given to the Infantry units in the vicinity, so that they may themselves undertake the service of security and reconnaissance in the narrowing zone between the Armies. In this case the Independent Cavalry Brigade had been absent all day, and the Cavalry Corps must have known this, and should have taken steps to take over the protection of the Right flank, even in default of orders to that effect. Had the Cavalry Corps been trying to execute its orders to turn Gumbinnen and advance on Insterburg, there might have been some excuse. But after the successful action against the Landwehr, the Cavalry Corps simply withdrew, without attempting to seize the crossings of the R. Inster and prepare for the advance next day. All touch was lost with what was presumably the German left flank, and no attempt was made to envelop it. The Corps had covered some 20 miles on this day.

The Independent Cavalry Brigade, still in bad communication with the Army, remained at Schillehnen during the 19th.

The right Division of the Advance came into contact during the morning with the advanced troops of the Ist German Corps on the line Gr. Kannapinnen–Niebudszen. They had blundered blindly into the enemy, which is not surprising when one reads the orders issued early that morning.

From these it appears that the Advanced Guard was not given any mounted troops whatever, and the Division was evidently moving into extended bivouacking areas, preparatory to a halt on the 20th. The movement of the Advanced Guard was further timed to commence upon the arrival of another body of troops at a given point. No indication of the probable time of arrival of this other body was given.

The Russians drove in the German Advanced troops, moving rapidly forward until held up by the prepared position Gr. Kannapinnen–Niebudszen. No attempt was made to throw out a flank guard on the right, and this flank was suddenly charged by German Cavalry and destroyed, the débris withdrawing to Mingstimmen. Over 500 Russians are buried at Uszballen, and others at Mingstimmen. Towards evening the 28th Division was badly mixed up, but managed to maintain itself on the line Uszballen–Brakupönen.

As an example of the attitude of a Russian Commander to the loss of men, the reports of the Commander of the XXth Corps to the Army are instructive. At 1 P.M. he reported, " I do not think that strong forces are moving against Lashkevitch (28th Division)," and later, " Evidently only dismounted Cavalry advancing against Lashkevitch, we are continuing our advance." At this time the 28th Division had suffered over 3,000 casualties, and was in much disorder. The country is ideal for observation, and there are several points from which the attack and rout of the 28th Division could have been observed with glasses. Neither the Divisional nor Corps Commander appears to have come up to reconnoitre, although it is possible to move anywhere mounted.

The fighting of the 28th Division is an example of how Infantry may become involved if they advance without proper direction from behind, consequent upon lack of reconnaissance.

It is well known how it is the nature of men to turn and face the fire which is directed upon them. In this case the Germans had cleverly refused their left flank, the village of Niebudszen being the angle, so to speak. The Russians, finding what appeared to be a flank, pressed more and more forward on their right until they almost faced South opposite Gr. Kannapinnen, and still found the Germans presenting a front. The moment soon arrived when there were no more troops either to extend the flank to the right or to protect it in case of attack. At this moment the German Corps Cavalry, which had been waiting behind the rising ground at Antbrakupönen, came over the hill and charged the 109th Regiment in flank and rear, driving it back in disorder. The whole line then fell back under heavy fire.

This refusal of the German flank was a species of trap into which the Russians blundered. Had there been a proper reconnaissance, the troops would have received an initial direction from behind, seizing such tactical features as were essential to the protection of front and flanks before continuing their advance.

The 29th Division, immediately South of the 28th Division, advanced in better order. Coming into contact with the German advanced troops during the morning, it drove them steadily back towards

Niebudszen–Springen. The Advance ceased when the effect of the disaster to the right flank of the 28th Division became felt, and the Division settled down for the night in close contact with the enemy.

The remaining troops reached their objective without fighting.

The Divisions settled down into extended bivouacking areas on the night of the 19th. There were no orders issued by the Army or by the various Corps for the holding of any line of resistance, in case of attack that night or during the halt on the 20th. Here was a glaring instance of bad Staff work, for such an order should be a matter of routine.

The 28th Division had withdrawn from close contact with the enemy after its disaster, but the 29th Division maintained its position. No attempt was made in either Division to throw out any continuous line of outposts. Units merely arranged for alarm posts to their immediate front.

Rennenkampf and his Corps Commanders seem to have been firmly convinced that the enemy was withdrawing hastily in front of the Russian advance, and that the opposition to the XXth Russian Corps was merely a rear-guard action. No orders were, however, issued to the Cavalry to verify this supposition and to keep touch with the enemy. The Cavalry were to rest on the 20th, equally with the remainder of the Army.

The position was, however, entirely different. On 14th August, Von Prittwitz had issued orders for a concentration against Rennenkampf.

8. German Orders. (Map II.)

German Eighth Army Order.

14th August 1914. 4 *p.m.*

1. It appears not unlikely that the enemy may advance South of the Forest of Rominten against the line of the Lakes. His advanced troops—Infantry and Artillery—are advancing from the line Gr. Czymachen–Mierunsken.

2. The Army will concentrate on its left flank to take the offensive against the enemy's advance.

3. The XXth Corps will advance to the neighbourhood of Ortelsburg. To this Corps falls the task of protecting the Army against any enemy advance South of Lake Spirding. An offensive against Johannisburg is to be kept in view, and the detachment at Bialla will therefore remain there for the moment. The 3rd Reserve Division, with the 6th Landwehr Brigade, will hold the line Nikolaiken–Lötzen. The Ist Reserve Corps will hold the line of the R. Angerapp, with its right wing on Lake Mauer. The XVIIth Corps will entrain *viâ* Allenstein and Königsberg for Insterburg–Norkitten, and thence by road to Darkehmen.

The Ist Corps will remain for the moment at Gumbinnen–Insterburg.

The 2nd Landwehr Brigade will hold the line of the Memel.

The covering of the right flank of the Army against the, in any case, very unlikely event of a Russian advance from the Narew is placed in the hands of General von Unger with the following detachments :—

(*a*) Garrison of Thorn at Strasburg.
(*b*) Garrison of Graudenz at Lautenburg.
(*c*) 70th Landwehr Brigade at Mlawa–Soldau.
(*d*) Garrison of Danzig at Neidenburg.

4. Reconnaissances will be carried out as follows :—

XXth Corps—Makow–Lomja–Osowiec.
3rd Reserve Division—Osowiec–Augustow.
1st Reserve Corps—Augustow–Suwalki–Wizajny.
I[st] Corps } Przerosl–Kalwarja.
1st Cavalry Division } Wylkowyszki.

Air Reconnaissances :—

XXth Corps—towards and beyond the Narew.
I[st] Corps—R. Niemen to Tilsit.

General Unger will issue his own instructions for reconnaissance of the line Ciechanow–Przasnysz.

5. Aviation Detachment 8 will be moved to Nordenburg, and will be placed at the disposal of the I[st] Reserve Corps.

6. Rear Communications :—

XXth Corps—Allenstein–Osterode.
3rd Reserve Division—Lötzen–Rastenburg–Sensburg.
I[st] Reserve Corps—Angerburg–Bartenstein–Bischofstein–Heilsberg.
XVIIth Corps—Darkehmen–Nordenburg–Gerdauen–Schippenbeil–Bartenstein.
I[st] Corps—Insterburg–Jodlauken–Muldzen–Allenburg–Friedland–Pr. Eylau.
Königsberg Garrison—Roads on each side of the R. Pregel to Königsberg.

7. Headquarters will remain till the evening of the 15th in Marienburg, after that Bartenstein, opening at 12 noon on the 16th.

V. Prittwitz, *Commander-in-Chief.*

9. Comments upon the German Orders.

The German Commander had decided to adopt a forward policy in East Prussia, and to deal with the First Russian Army, which was the more menacing for the moment. He had decided to take the offensive, and was concentrating on his left flank ready for a suitable opportunity.

The grouping is excellent. One Corps only is left to meet any movement on the part of the Second Russian Army. Garrison Detachments from the Vistula are left to protect the right of this Corps against any advance towards Soldau. A weak Landwehr Brigade, supported for the moment by a Reserve Division, is left to guard the Lötzen Gap. The whole of the remainder of the Eighth Army is moved up ready for the intended offensive. There are no half-measures about this order.

It should be noticed that the I[st] German Corps is ordered to remain about Gumbinnen–Insterburg.

Everything seemed to point to a battle with the First Russian Army on the line of the R. Angerapp. It was in the interest of the general strategy of the campaign not to fight a battle too far to the East, for every march farther Eastwards meant further dispersion from the forces left to face the Second Russian Army. As it was intended to return to deal with the Second Army after the First Army had been defeated, this was a matter of importance. It was actually in favour of the Germans to allow the Russians to advance Westwards into East Prussia, despite the disadvantages of deliberately permitting an Invasion with all its horrors.

Von François, the Commander of the Ist Corps, which hailed from East Prussia, construed his orders into a more forward defence than had been intended by Von Prittwitz. As we know, he had located most of his Corps on a line with Stallupönen, considerably in advance of Gumbinnen. His fiery spirit could not tolerate an Invasion of sacred East Prussian soil, and his contempt for the Russians led him to adopt the offensive action which he had carried to a successful conclusion. The fighting at Stallupönen was a complete surprise to the Headquarters of the Eighth Army, and considerably shook their *moral.* They feared, very properly, that the Eighth German Army might be forced to move too far Eastwards to support the Ist Corps, and so bring on prematurely the decisive battle that was intended.

Von François was undoubtedly wrong in pushing so far forward, but the situation brings out clearly the difficulties of ordering a deliberate abandonment of national territory without a struggle. The battle of wills between Von Prittwitz and Von François undoubtedly created difficulties for the Staff directing the German Campaign, and must have weakened the Army Commander's confidence in his subordinate's desire to carry out his orders. Von François could equally well have organised his excellent Frontier protection farther back from the Frontier without any alteration in tactics.

10. German Situation on 19th August. (Sketch C.)

By mid-day on the 19th the position of the German troops was as follows :—

6th Landwehr Brigade—Lötzen.
3rd Reserve Division—Angerburg.
I[st] Reserve Corps—R. Angerapp (from Angerburg to Darkehmen).
XVIIth Corps—Concentrating on Darkehmen.

<table>
<tr><td>I[st] Corps
1st Cavalry Division
Königsberg Garrison</td><td>In front of Gumbinnen under the I[st] Corps Commander.</td></tr>
</table>

2nd Landwehr Brigade—Kraupischken.

The German plan had taken definite shape. The position of the R. Angerapp was being strongly occupied. The line of the Lakes was being left to one Landwehr Brigade only, and the whole of the remaining forces were being made available for active operations. The right flank was secure against the Lakes at Angerburg, and the left flank was well thrown forward to ascertain the extent to the North of the Russian Advance. Could the German Commander entice the Russians to attack him in this prepared position, he would have all the conditions in his favour. The direction of the enemy advance was now well-known, and a battle might be expected about the 21st at the latest. During the early hours of the 19th the Russian order for the 19th and 20th was picked up by the German wireless. It was then found that the Russians intended to halt on the 20th. The first indications also of a Russian advance on the Narew had also begun to come in. Samsonoff's Army was on the move unexpectedly. What was the German Commander to do now ? How long did the First Russian Army intend to halt ? Had the Germans time to await the Russians in the position of the R. Angerapp, or would it be necessary to advance against them, giving up all the advantages of a prepared position, against which the Russians could be forced to expend their strength, before being dealt with by a counter-stroke ? The timing of such an operation required careful consideration.

The personality of the German Commander was to decide the question in a large measure. Von Prittwitz was weak, and his Ist Corps Commander, Von François, was strong.

Von François had taken up his position in front of Gumbinnen with the curious left flank thrown back to Gr. Kannapinnen. The 1st Division was holding the left, with Von Brodrück's Königsberg Garrison on the right, extending a short distance South of the Stallupönen–Gumbinnen railway line. He had withdrawn the 2nd Division into reserve near Waivern and Bibehlen. The 1st Cavalry Division had also been ordered to bivouac near the 2nd Division.

The terrain in this part of the country is excellent, slightly undulating, and covered with stubble at the time of the action. The position enclosed the high ground with a long sloping field of fire in front of it, almost devoid of cover, if one excepts the isolated small farms and villages with their patches of trees.

Mallwischken overlooks Uszballen and Gr. Kannapinnen, and should have been the key of the Russian right, but they made no attempt to include it in their line after the disaster to the 109th Regiment. Von François evidently considered it too far off to include in his line, hence the thrown back flank, leaving the valley in front of him at Gr. Kannapinnen.

About mid-day on the 19th, Von François rang up the Army Commander and told him that he expected to be attacked that evening, or at least on the 20th, as he was still in close touch with the enemy. He told Von Prittwitz that the Russians were advancing carelessly, and that there was a golden opportunity for attacking them. It was well-known that the fiery Von François was violently opposed to giving up more of the sacred territory of East Prussia, and he impressed Von Prittwitz by his reports of the dreadful sight of the retreating refugees leaving the country wholesale. Von François begged to be allowed to attack, and for the remainder of the Army to come up level with him.

The insistence of Von François, combined with the effect of the initial success at Stallupönen and the possibility of a long halt on the part of the Russians in front of him, caused Von Prittwitz to yield. He issued orders for the Army to concentrate in line with the I[st] Corps. He informed Von François at 5 P.M. on the 19th that the remainder of the Army was to come up on his right on the line Gawaiten–Augstupönen.

Von François then issued his orders for the attack :—

2nd Division—From Mallwischken on Mingstimmen at 4 A.M.

1st Division—To open a heavy Artillery fire at 3.30 A.M., and to be ready to co-operate as soon as the 2nd Division attack was seen to be under way.

V. Brodrück's Königsberg Garrison—To remain in position ready to attack.

2nd Landwehr Brigade—To occupy the crossings of the R. Inster, and prevent the Russian Cavalry from crossing near Kraupischken.

1st Cavalry Division—To follow the 2nd Division ready to attack.

Each of these units was supplied with liaison officers from the Ist Corps Headquarters, which established its position at Lindenkrug on the main Gumbinnen–Mallwischken road. From here a careful system of despatch-riders was arranged.

The Corps Commander says that he had a view of the battlefield from his table placed opposite the inn at Lindenkrug, but this is not so. He must have been well forward of this point, near Blecken, if he actually did have a view of the battle, or just South of Gr. Kannapinnen if he saw the 2nd Division attack.

11. The Events of 20th August. (Sketch C.)

It was still hardly light when at 3.30 A.M. the Artillery of the 1st Division, strengthened with the Corps Heavy Artillery, opened fire. The Russians were caught completely unawares.

It seems almost incredible that the XXth Russian Corps (28th and 29th Divisions) should have deliberately settled down to rest on the 20th, when they had been fighting at close quarters on the 19th, and had actually been held up and one Division so roughly handled that it had been compelled to retire. But they had certainly done so, and that without making sure of the high ground at Mallwischken, which lay unoccupied by either side during the night of the 19th/20th. The Commander of the XXth Corps could not possibly have made a personal reconnaissance or he would have realised the position. Corps Headquarters were at Kattenau, only 6,000 yards behind the front line, with the 28th Division Headquarters at Mingstimmen.

The German 2nd Division commenced its advance from Sassupönen to the minute at 4 P.M. They had made a night march of 10 miles through the Forest *viâ* Wilpischken and Rohrfeld between 9 P.M. on the 19th and 4 A.M. on the 20th. The road is rough and unmade, and this performance of a whole Division with its Artillery must be regarded as a fine one, especially as it was followed by a successful attack over a farther 7 miles.

Uszballen was reached at 7 A.M., and the Russians driven helter-skelter Eastwards. From Mallwischken the Artillery of the 2nd Division must have had some magnificent targets against the Russians withdrawing in disorder across the valley and up the slopes towards Tutschen–Kalbassen.

At 8 A.M. Von François launched the 1st Division, swinging it round to attack the left of the 28th Russian Division at Brakupönen. The timing of this attack must have been comparatively easy, since the whole of the 2nd Division's attack passed in full view of the line Gr. Kannapinnen–Niebudszen.

By 12.30 P.M. the 28th Russian Division was trying to reform East of Tutschen–Kalbassen, having suffered heavy casualties, and having lost 5,000 prisoners. Von François says that his *moral* was greatly raised by seeing such a large number file past his Headquarters.

The I[st] Corps stopped for breath on the line Brakupönen–Bruszen, with the Corps Cavalry in pursuit of the Russians.

Von Brodrück's Königsberg Garrison, feeling the troops on its left advancing and seeing some of the Russians on its left front retiring, decided to commence the advance without further orders. Von François had not intended this to take place until the attack on the left had more fully developed, and the expected XVIIth Corps attack came up on his right.

He doubted the capacity of the Landwehr to advance against the Russian First-Line troops. But Von Brodrück seems to have listened to the incitements of the Liaison Officer, a young Lieutenant from I[st] Corps Headquarters. The advance did not go far, as the Landwehr units suffered some casualties from their own Artillery, finally settling down again in their old positions.

Here is an example of a Commander keeping things very closely in his own hand owing to the proximity of his Headquarters to the fighting, and yet finding things getting out of hand. He would have done better to have disclosed his intentions verbally to Von Brodrück, or in the form of special Instructions, which could be referred to in moments of doubt.

The situation on its right, combined with the Landwehr attack, caused the Russian 29th Division to withdraw to the line Kummeln–Puspern–Kl Puspern, suffering some casualties as they withdrew across the open valley.

The 1st German Cavalry Division, which should have spent the night near the 2nd Division, had moved up Northwards to meet the Russian Cavalry Corps, and never received its orders. Finding no Russian Cavalry on the evening of the 19th, for as we know it had withdrawn to Lindental after the successful attack against the Landwehr at Kauschen, the German Cavalry moved on towards Spullen as soon as it was light on the 20th, and the sound of the guns farther South became audible. Here they found Oranofski with his Independent Cavalry Brigade, having just arrived. Spullen is an outstanding hill from which a view of the whole of the ground over which the I^st German Corps attacked can be obtained. A few shells were sufficient to cause Oranofski to ride off the field. He retired to Schillenen again, losing touch with the Army for several days. The German Cavalry then moved down to Schwirgallen, charging the Russian transport, causing a panic, and completing the disaster of the 28th Russian Division.

This is a fine example of a Commander, temporarily without orders, moving to the sound of the guns and taking an active and intelligent part in a battle he finds in progress. In direct contrast was the action of the Russian Cavalry Corps which remained inactive at Lindental, although the sounds of a severe battle must have been distinctly audible.

The incident also shows the impotence of Cavalry unaccompanied by guns, or acting far from the support of Infantry with its Artillery. Oranofski had no Artillery to oppose to the 3 Horse Batteries of the 1st German Cavalry Division.

All had gone well with the Germans in the North, but matters were much otherwise in the South.

The XVIIth Corps debouched across the R. Angerapp in two columns; the 36th Division from Darkehmen towards Grünweitschen, where the left of a Russian Corps was supposed to rest; and the 35th Division from Kissehlen on Szirgupönen against the left flank of the Russians opposing Von Brodrück. The movement of both Divisions commenced at 5.30 P.M. on the 19th, and continued throughout the night. Administrative arrangements for the night march were bad, and there was much halting and confusion. Nothing had been done to clear the roads of the masses of fugitives fleeing in front of the Russians. Vehicles and live-stock blocked the roads everywhere, and the troops had to be diverted on to side roads on several occasions. Irrespective of marches in the forenoon of the 19th, all units of the 36th Division were on the move from 5.30 P.M. on the 19th till they came into action at 8 A.M. the next morning. During this 14½ hours they had covered some 18 miles. The attack of the XVIIth Corps, coming as it did some four hours after that of the I[st] Corps, was no surprise, and the Russians were ready to meet it.

On the right the 36th Division drove in the enemy outposts at Schwiegseln and Grünweitschen, but was brought to a standstill by the Russian Artillery fire just to the East of these two villages. Only on the extreme right did they manage to push forward some distance to Kubillen.

On the left the 35th Division made some progress, reaching and passing the right of Von Brodrück's line. Had they come up at the time that the latter made his ineffective attempt to advance, they might have influenced the battle, but were eventually forced to withdraw, after suffering severe casualties, behind the Schwentischke stream.

Von Mackensen appears to have thrown in both Divisions piecemeal under the impression that the Russians were withdrawing. Up to noon on the 20th his reports were optimistic, but, unsupported by their Artillery, the German Infantry suffered much, and there was a panic withdrawal on more than one occasion. All the Superior Commanders, including Von Mackensen himself, became involved in stemming the rout, which had assumed alarming proportions. By evening the Corps was disposed on the line Kubillen–Schwiegseln–Gr. Baitschen, and thoroughly exhausted. They had been opposed by three Russian Divisions—the 25th, 27th, and 40th, —which had had time to leave their extended bivouacs and form up in good order to meet the German advance.

The Ist Reserve Corps had debouched from Dombrowken and to the South of it. They also had had a trying night march, since considerable difficulty had been experienced in collecting the detachments spread defensively along the R. Angerapp. It was not till 11 A.M. on the 20th that the Advanced Guards of the 1st Reserve and 36th Reserve Divisions reached Kleszowen and Friedrichsberg respectively. The Corps Cavalry were at fault in their reconnaissance, for they reported Goldap and the roads leading from it free of the enemy. The Corps Commander directed his divisions towards the N.E., with the object of taking the Russians opposing the XVIIth Corps in flank, but no sooner had the movement commenced than strong forces were reported advancing from Goldap and Gawaiten. Front had to be made to these, and by evening the Corps was engaged along the line Wilkatschen–Kurnehnen–Grischkehmen with considerable Russian forces. The men were very tired, though they had not had much fighting. The Corps had not been able to give any support to the XVIIth Corps.

The 3rd Reserve Division, which had been at Angerburg, was far behind, and only succeeded in reaching Benkheim by the evening of the 20th.

12. The German Withdrawal.

At 6 p.m. on the 20th Von Prittwitz rang up Von François, and told him that neither of the other two Corps were doing well, and that the news from the XXth Corps opposite Samsonoff was bad. He added, "I must withdraw to the Vistula."

Speaking to Von Mackensen, he said, "The situation has completely altered. New forces have appeared in the North. The whereabouts of the 1st Cavalry Division is not known. The I[st] Reserve Corps is opposed by equal forces. Fresh forces of about an Army Corps have appeared about Lyck.

The Narew Army is advancing in strength against Ortelsburg and Soldau. The Eighth Army is therefore going behind the Vistula."

Von François protested vigorously, pointing out that the situation was favourable for an attack the next day, but about 10 P.M. he received the final order to withdraw.

The order for withdrawal to the Vistula was apparently issued telegraphically about 9.30 P.M. on the 20th.

13. Comments on the German Operations.

Von Prittwitz had undoubtedly adopted the correct course in moving forward to the R. Angerapp to oppose the First Russian Army. He had concentrated every available man for this purpose. The unexpected halt of Rennenkampf, as ordered for the 20th, had then upset his calculations for awaiting him in his selected position. He had allowed himself to be hurried into action by one of his subordinates. He could not possibly have concentrated all his troops for a battle in line with Von François on the 20th. The problems of time and space must have been wrongly worked out by the Staff of the Eighth German Army, or what is more likely, Von Prittwitz, having weakly yielded to the insistence of Von François, overruled the objections of his Staff. He allowed his three Corps to attack in succession at 4 A.M., 8 A.M., and 12 noon respectively, while the 3rd Reserve Division on his right was not released from Angerburg till noon on the 20th, and never reached the battlefield. He therefore literally did not control the fighting.

The contrast between the results achieved by the I[st] Corps on the one hand, and the XVIIth and I[st] Reserve Corps on the other, clearly exemplifies the difference between a prepared and an unprepared attack. With the I[st] Corps all was ready. The operation took place over ground which had been well reconnoitred, and was known to subordinate Commanders and troops. In the case of the other two Corps, everything was hurried. The men were thrown piecemeal into the fighting without reconnaissance, and without proper Artillery support.

Had proper calculations been made as soon as the Russian intention to halt had been discovered, Von Prittwitz would have reasoned the problem as follows: Were the Russians likely to make a prolonged halt in front of him on the R. Angerapp, and so force him to face about and march to oppose the Narew Army, or had he sufficient time in hand to continue to await them on the R. Angerapp position? If he marched forward to meet the Russians in front of the R. Angerapp he was abandoning the advantages of a prepared position, but, on the other hand, the sudden assumption of the offensive might come as a surprise. If he marched forward, the earliest date upon which he could concentrate in line with Von François was the 21st. On that date he might find the Russians advancing once more.

Von Prittwitz's mistake was in moving forward and attempting at the same time to fight on the 20th.

The situation about 6 P.M. on the 20th at Eighth

German Army Headquarters presents an interesting psychological study in the art of command. Von Prittwitz had started his operations in a high state of optimism. The Second Russian Army did not seem ready to start. Despite the unruliness of Von François all had gone well. The Russians had suffered heavily in the fighting of the 17th and 19th. The German troops had fought well. On the evening of the 19th the die had been cast for an offensive. There must have been a certain sense of relief. On the 20th Von François had done magnificently, and up till 12 noon all reports were favourable.

Then quickly the picture began to change. On the left Von François had done all he could for the day. In the centre Von Mackensen's earlier rosy reports began to darken a little. He reported at first a check, and finally that he did not feel himself equal to withstanding a Russian attack should such develop. On the right the Ist Reserve Corps had been late in coming up, and had found no flank to turn. The 3rd Reserve Division had not reached the battlefield.

It must have been evident that there had been miscalculations in the concentration marches. Decisive success could not be achieved that day.

No news could be obtained of the 1st Cavalry Division. Had it been swallowed up by the mass of Russian Cavalry ? Was the Russian Cavalry even now penetrating far in rear of the German left flank ?

And then to cap all came the definite and unexpected news that the Second Russian Army was not only on the move, but that it was close to the Frontier opposite the line Ortelsburg–Soldau. It was in great strength, and extended farther westward than had been expected.

Under the weight of all this bad news the optimism of the Commander-in-Chief collapsed entirely. In his Chief of Staff, Graf von Waldersee, he found an only too ready participant of his fears.

General Hoffman, at the moment the General Staff Officer in charge of operations with Von Prittwitz, has given us in his book an account of the scene which occurred after the receipt of the news about the Second Russian Army. It is worth repeating, for it gives a picture of the conditions prevailing at German Headquarters.

Hoffman and Grünert, the Quartermaster-General, were discussing the news. Hoffman remarked, "I fear that the nerves of the Commander-in-Chief and of the Chief of the Staff will not be equal to this report; I should prefer to suppress it. We could then finish off the fight here to-morrow, and afterwards turn on our Warsaw enemy." Grünert replied, "You could not withhold such an important report from the Chief." Although not spoken in earnest, the conversation portrays the lack of confidence the Staff had in their superiors.

At this moment Von Prittwitz appeared, and Hoffman and Grünert were called into the Commander-in-Chief's room. Von Prittwitz told them that he had decided to go behind the Vistula. Both Hoffman and Grünert protested, pointing out that there was still time to defeat Rennenkampf. Von Prittwitz silenced them curtly, telling them that the responsibility of decision rested with the Commander-in-Chief and not with them.

Von Prittwitz was then left alone. As we know, he had already communicated his decision to both Von François and Mackensen. He now called up Supreme Headquarters at Coblentz, and told them of his decision. His news must have been a veritable bombshell. He painted a very dark picture of affairs in East Prussia, and even stated that he was doubtful of being able to check the Russians on the Vistula, owing to the lowness of the water in the river.

Hoffman and Grünert then continued the argument with the Chief of Staff, Von Waldersee. Hoffman pointed out that a withdrawal behind the Vistula could not be carried out without fighting. The troops had to be disengaged from contact with Rennenkampf's Army, and the left flank of Samsonoff's Army must be checked, since it was closer to the Vistula than the main body of the Eighth German Army. He suggested concentration for an offensive against Samsonoff's left flank in order to carry out the withdrawal to the Vistula. Von Waldersee agreed.

On the evening of the 20th, orders were sent out in accordance with this plan. Hoffman gives these orders in substance as follows (Map II.):—

(i) The XXth Corps to concentrate near Hohenstein.
(ii) The I[st] Corps and 3rd Reserve Division to entrain for transportation towards the right flank of the XXth Corps.

(iii) The Königsberg Garrison to cover the entrainment and then withdraw to the Königsberg Area.

(iv) The I[st] Reserve and XVIIth Corps to withdraw Westwards, disengaging from Rennenkampf.

From the above account, which seems to be substantiated by both Russian and German accounts, it will be seen that considerable confusion reigned at Eighth Army Headquarters. Von Prittwitz had personally informed Supreme Headquarters and two of his Corps Commanders of his intention to go behind the Vistula. He had left his Staff under the impression that orders were being issued to that effect. He did not worry himself to ascertain whether these orders had been issued. No one informed Supreme Headquarters that there had been any change of plan.

Von Prittwitz was a beaten man.

Having once put his hand to the plough there should have been no turning back. He should have persisted until all hope of defeating Rennenkampf had disappeared. Half measures were of no use in such a position as his.

And, indeed, the position was by no means so serious as he regarded it. The Narew Army undoubtedly menaced his communications through Allenstein, but the delaying power of a force of about two Corps was not to be despised, and the Russians were advancing over a very difficult country, where their advance must necessarily be slow. He had also his Northern line to the Vistula through Königsberg, which was not menaced by the Russians. One of his Corps Commanders, Von François, was still optimistic, and the reports of Von Mackensen might be discounted somewhat from the fact that he had been personally involved in an unfortunate panic on the battlefield, and had been adversely affected. The troops had been exhausted by a long night march, but the power of recuperation of good troops, after even a few hours' rest, is great. On the right there were still troops which had not been engaged.

In actual fact, from the table of troops engaged at Gumbinnen, it will be seen that the Russians were actually inferior to the Germans in all but Cavalry. It was not to be supposed that they had not suffered heavily in the fighting of the 17th/19th and 20th, a fact which a weak commander appears to forget when reckoning up his own casualties. The Germans were superior in Infantry by 14 Battalions, the proportion of Infantry Battalions which actually took part in the fighting being Germans, 83 Battalions; Russians, 78 Battalions. In Artillery the Germans were greatly superior, especially in heavy Artillery, the proportion of guns actually employed being Germans 390, Russians 330.

In any case, however, the decision to go behind the Vistula immediately was unjustifiable. He gave his Staff no idea of how such an extensive operation was to be carried out.

Such a Commander was of no use to Supreme Headquarters. He had failed in the trust which had been placed upon him. On the night of the 21st/22nd it was decided to supersede him and his Chief of Staff. The new choice fell upon men of very different stamp.

14. Comments on the Russian Operations.

Rennenkampf had been caught napping, having ordered a halt for the 20th without reference to the situation of the enemy, whose positions he had failed to reconnoitre. The reckless use of wireless had, moreover, informed the enemy of his intention to halt. He apparently considered it a foregone conclusion that the enemy would retreat in front of him, and would make no attempt to fight a battle. From notes made by one of his Staff, it would appear that Rennenkampf's advance was intentionally slow, irrespective of his desire to organise his rear services. He knew that his rôle was to draw upon himself the greatest possible strength and to entice the enemy Eastwards, so that Samsonoff's Army might fall upon his rear. Knowing that Samsonoff's Army might fall behind its timing, he deliberately delayed his march. But there can be no excuse for not making definite arrangements for the holding of positions should the enemy suddenly decide to attack. Such arrangements should be in the nature of ordinary Staff routine.

The Germans surprised Rennenkampf, and with no preparations for defence he never regained control during the 20th. The Battle was fought out by the Russian Divisions acting as each thought best.

Rennenkampf's system of issuing daily Instructions to the Cavalry fell to the ground. It is also inexplicable that he should have ordered the Cavalry to halt equally with the Infantry on the 20th. The Cavalry Corps had only covered 48 miles in three days (17th, 18th, and 19th), and their transport was not incomplete as was that of the Infantry. Reconnaissance to a distance in front is all the more necessary in the case of large forces intending to halt.

Even without orders the inaction of the Russian Cavalry leaders is inexcusable. Hearing the sound of battle on the 20th, the Russian Cavalry Corps might have been expected to take an active part. It might have moved Southwards towards the sound of the guns, or it might have continued to move on Insterburg in accordance with its original orders. Had the Cavalry Corps moved South from Lindental, it would have prevented the 1st German Cavalry Division from attacking at Schwirgallen, and it might even have taken the 2nd German Division in flank, thus saving the 28th Russian Division from so complete a disaster. Had it moved on Insterburg, it is not to be expected that the Landwehr Brigade, already heavily punished on the 19th, would have delayed their advance for long, considering that the R. Inster is only some 20 feet wide at this point, and in the summer only some inches deep. The arrival of a fighting force of 14,000 sabres and 48 guns at Insterburg might well have spread panic in the German rear, and turned their withdrawal on the 21st into a disaster. The Commander of the Cavalry Corps, Khan Nachivanski, appears to have been quite unsuited for his position. Oranofski had been without orders since the campaign began. He appears to have acted almost in a cowardly manner when he found himself under gunfire and unable to reply. He employed his mobility in disappearing from the scene.

Gourko, on the left flank, was a man of another stamp, who afterwards rose to high rank. His inaction on the 20th appears to have been due to his orders to protect the Army against an attack from the direction of Lötzen. He had thrown out several strong reconnaissances in this direction, from which he had received no information, and he hesitated to move Northwards with the residue of his Cavalry Division. On the night of the 19th the 5th Rifle Brigade had been taken from him, and placed under the Commander of the IVth Corps. The counter-order replacing it under his command only reached the IVth Corps and not Gourko. The Commander of the 5th Rifle Brigade did not know under whose orders he was, and receiving none, decided to halt quietly in bivouac and rest.

15. Results of the Battle of Gumbinnen.

The Battle of Gumbinnen was a strategical success for the Russians, in that the Germans withdrew from the field after having intended to fight a decisive action. The confidence of German Supreme Headquarters was rudely upset. The change of command was a difficult matter to settle, and Von Moltke must have had his mind diverted from the main theatre of war in France by the fear of not having sufficient troops in the East. One of his big Commanders had failed under the pressure of modern war conditions; some of the others might do so also. Were Hindenberg and Ludendorff the right men? Were the German troops after all not superior to the Russians in fighting value? Such thoughts as these must have clouded the judgment of the leader of the German Armies in some degree.

The German casualties were heavy, though actual figures are not available.

The Russian official casualties on the 20th August are given as follows :—

28th Division . . .	104 officers	6,945 men
29th "	22 "	3,000 "
25th "	35 "	3,145 "
27th "	21 "	950 "
40th "	31 "	2,022 "
30th "	10 "	800 "
Total . .	223 officers	16,862 men

This loss may be taken as being about 20 per cent of the Infantry engaged. One-third of these casualties were due to prisoners, chiefly in the 28th Division.

16. The Forces Engaged.

The following table gives the troops available on both sides at Gumbinnen :—

BATTLE OF GUMBINNEN, 20th AUGUST 1914.

Russians.

	Battns.	Sqdns.	Guns. Field.	Guns. Heavy.
*Cavalry Corps .	..	72	42	..
*Oranofski's Brigade	..	12	..	..
*Gourko's Division .	..	22	12	..
*5th Rifle Brigade .	8	2	24	..
XXth Corps . .	22	3	114	..
IIIrd Corps . .	27	9	108	..
IVth Corps . .	29	2	108	..
1st Heavy Arty. Div.	..	..	..	12
	86	122	408	12

*On the Russian side—
8 Battalions Infantry, 116 Squadrons Cavalry, 78 Field Guns } took no part in the battle.

Germans.

	Battns.	Sqdns.	Guns. Field.	Guns. Heavy.
Ist Corps . .	24	8	144	16
Von Brodrück .	11	7	12	12
XVIIth Corps .	24	8	144	16
Ist Reserve Corps .	24	8	72	..
*2nd Landwehr Bde.	5	5	12	..
1st Cavalry Division	..	24	18	..
*3rd Reserve Division	12	4	72	..
	100	64	474	44

*On the German side—
17 Battalions Infantry, 9 Squadrons Cavalry, 84 Field Guns } took no part in the battle.

SECOND ARMY OPERATIONS.

1. Samsonoff's Orders.

Samsonoff arrived at Bialystok from the Caucasus on 12th August. On 13th August he received Jilinsky's Instruction. That same day his Chief of Staff prepared a Memorandum for him showing how it was proposed to concentrate the Second Army in accordance with this Instruction.

The Memorandum ran as follows :—

"In accordance with the instructions of the Commander of the North-West Front, dated 13th August 1914, the Second Army is to advance from the line Augustow–Grajewo–Myszyniec–Chorzele to the line Rudczanny–Passenheim, and farther to the North on the flanks and rear of the line of Lakes.

In addition, the Second Army is ordered to push Cavalry across the frontier on the whole front, supported by Infantry where possible, on the 18th August, followed by the main bodies of Corps on the 19th August.

The neighbouring First Army, advancing from the front Wierzbolowo–Suwalki, is to cross the frontier on the 17th August, directing the left flank Corps from Suwalki on Angerburg.

The most suitable areas of concentration would appear to be :—

IInd Corps—Augustow.
VIth ,, —Lomja.
XIIIth ,, —Ostrolenka.
XVth ,, —Rozan.
XXIIIrd ,, —Novo Georgievsk.

IInd Corps.

In the near future all the IInd Corps will be concentrated in the Augustow area, with the exception of one Brigade of the 26th Division and 24 Guns in Grodno.

In view of the necessity of protecting the right flank of the Second Army against the exits from the front Lötzen–Johannisburg, the IInd Corps is to leave Augustow on the 17th.

VIth Corps.

The VIth Corps is located on the Bobr from Dolistowo to Wizna.

In accordance with the Instructions of the Commander of the North-West Front, the main bodies of Corps are to cross the Frontier on the 19th.

Therefore the VIth Corps must commence its advance from Lomja on the 18th.

To effect this it is necessary that the Corps should be concentrated by the evening of the 15th towards its left flank, and thence proceed to the Lomja area (27 miles in two marches).

XIIIth Corps.

This Corps, in accordance with a telegram from the Army, has already begun its movement from Bialystok to Ostrolenka, and by the evening of the 17th will be in its assigned position.

Thus this Corps can advance on the 18th, though it should be mentioned that in its movement to Ostrolenka the Corps will be called upon to complete four lengthy marches (from 20 to 23 miles per day), with one halt in addition.

XVth Corps.

This Corps, now occupying Zambrowo, can reach Rozan on the 16th. It can therefore advance on the 18th from that place. Zambrowo–Rozan is 45 miles.

XXIIIrd Corps.

As has already been stated, this Corps was broken up at the commencement of mobilisation, until only the Corps Staff remained.

In accordance with orders from the Commander of the North-West Front, the 2nd Division is to be replaced at Novo Georgievsk by the 1st Rifle Brigade.

The 3rd Guard Division must move from Bialystok to Lapy, entrain there for Wyzkow on the 14th, detraining on the 17th. On the 18th it can move to Pultusk (against this the Army Commander wrote, 'Leave at Bialystok'), and at the same time the 2nd Division can move to Nasielsk.

Second-Line Divisions it is proposed to employ as follows :—

76th—Grodno.
59th—2 Regiments Bialystok.
2 Regiments Osowiec.
Relieving 2 Regiments of the 16th Division (VIth Corps).
79th—Novo Georgievsk.
Relieving the 1st Rifle Brigade.
77th—Warsaw.

Thus the 1st Rifle Brigade can be moved up to follow the XXIIIrd Corps.

The Cavalry of the Second Army (4th, 6th, and 15th Divisions) has been given the task of seizing important points and crossings, and of raiding far in rear of the enemy, in order to prevent the removal of rolling-stock.

As the Main Force of the Second Army had been ordered to advance on the front Rudczanny–Passenheim and farther to the North towards Rastenburg, it would appear to be necessary to move up the 4th Cavalry Division from Szczuczyn to the front of VIth and XXIIIrd Corps (against which the Army Commander wrote, 'The 4th Cavalry Division is to remain at Szczuczyn ')."

On the night of the 14th August, the following Army order was sent out telegraphically :—

Second Army Order.

" Corps have been ordered to take up the following positions by 6 P.M. on the 17th August :—

IInd Corps—Augustow.
VIth ,, Lomja.
XIIIth ,, Ostrolenka.
XVth ,, Rozan.
XXIIIrd ,, —(less 3rd Guard Division and 1st Rifle Brigade) — Novo Georgievsk.
4th Cavalry Division—Szczuczyn.
6th and 15th Cavalry Divisions—Area of Ciechanow."

On the 16th August, Second Army Instruction No. 1 was issued from Bialystok :—

Second Army Instruction No. 1.

" The enemy in front of the Second Army has not changed except near Gutten (near Johannisburg), where about a Division of the XXth Corps has appeared, and about a Division of the XVIIth Corps at Mlawa. Neidenburg is being strongly fortified.

The Second Army, reinforced by the Guard and the I[st] Corps, will advance to the front Lötzen–Rudczanny - Ortelsburg - Passenheim, directing its main blow round the Masurian Lakes to the West, with the object of crushing the enemy operating in the Lake region, and cutting him off from the Vistula.

The advance of Corps will begin on the 18th, with the exception of the I[st] Corps, which will advance on the 17th.

The XXIIIrd Corps will move without the 3rd Guard Division and the 1st Rifle Brigade by road from Novo Georgievsk *viâ* Golymin–Opinogora–Grudusk–Kuklin, reaching Murawy on the 20th.

The VIth, XIIIth, XVth, and I[st] Corps will advance as follows :—

18th—Dobrylas–Zbojna–Grabowek–Ruzieck–Makow–Pomaski.

19th—Pupkowizna–Lipniki–Wola–Blendowo–Jednorozec–Przasnysz.

20th—Dombrowy–Pelty–Montwitz–Chorzele–Krzynowloga–Zaboklik–Mlawa.

The I[st] Corps advancing *viâ* the Nasielsk–Ciechanow–Mlawa road, and to the west of it.

The Guard Corps will concentrate in the Novo Georgievsk area.

The 4th Cavalry Division will reconnoitre between VIth and IInd Corps.

The 6th Cavalry Division will protect the left flank of the Army, reconnoitring to the left of the line Przasnysz–Janowo, and to the right of the high-road from Novo Georgievsk to Mlawa.

The 15th Cavalry Division will reconnoitre to the left of this high-road, and to the right of the line Plonsk–Drobin–Biezun–Strasburg.

The Headquarters of the Army will move to Ostrolenka on the morning of the 18th."

2. Comments on Samsonoff's Orders.

Samsonoff had been set a more formidable task than Rennenkampf. Rennenkampf was the Commander of the Vilna Military District, from which his troops came. All his movements for concentration had followed a long pre-arranged plan. Samsonoff, on the other hand, was the Governor of Turkestan. At the outbreak of hostilities he was on sick leave in the Caucasus, and hurried at once to Bialystok. He probably commenced the campaign in ill-health. He knew neither his Staff nor the troops he was to command. The concentration of his Army was being altered at the eleventh hour. Individual orders had been issued by the Commander of the North-West Front, but a complete Instruction was not received until the evening of the 13th August. A preliminary warning order had been issued on the 14th August, but the final Instruction of the Second Army was not issued till the 16th August.

The Advance across the Frontier, many days' march from the detraining stations of Corps, was to commence on the 19th August, whilst advanced troops had to cross on the 18th August. It will be seen how little time had been given for the issue and absorption of subordinate orders.

The whole movement was hurried. The calculations of time and space are a good example of the higher Staff work of the Russian Army. Had Samsonoff been in good health, or had he been an efficient Commander, he would have refused to accept the proposed movements of his Chief of Staff, and would have represented to Jilinsky the impossibility of the timing he was ordering.

The VIth Corps was on the right of the Main Advance. From Lomja to Dombrowy, the point where the right of this Corps was to cross the Frontier, is 35 miles. From Dolistowo, the right of the Corps concentration, to Lomja is 53 miles. The Corps was therefore to march 88 miles in five days (15th to 19th August), reaching the Frontier on the 19th, but not crossing it. Rate of marching was thus to be 17½ miles per day.

The XIIIth Corps was ordered to move off from Bialystok by road to Ostrolenka as units detrained, without waiting for rear services. From Bialystok to Ostrolenka is 84 miles. From Ostrolenka to the Frontier is 31 miles. To arrive on the 19th, therefore, the Corps had to cover 115 miles in seven days—that is, at a rate of 16½ miles per day without halting.

The XVth Corps was ordered from Zambrowo to Rozan, a distance of 41 miles, to be completed in two days (15th and 16th). From Rozan to the Frontier is 40 miles. To arrive on the 19th, therefore, the Corps had to cover 81 miles in five days—that is, 16 miles per day.

These distances were preliminary to a long Advance. They were to be carried out, over the worst possible roads, in very hot weather and with much of the transport deficient.

Such demands upon newly mobilised troops are bound to lead to many casualties from over-exertion.

Omitting the IInd Corps, which was to operate East of the Masurian Lakes, Samsonoff was advancing $4\frac{1}{2}$ Corps (VIth, XIIIth, XVth, half XXIIIrd, and Ist Corps) on a front of 47 miles. He had some excuse for making this wide extension, since the roads were few and bad, and there was no immediate probability of meeting the enemy. But it will be at once apparent that Samsonoff was directing his Advance far to the West of the objectives that Jilinsky had given him, and farther from the objectives of the First Russian Army. Jilinsky had ordered the Second Army to direct itself on the line Rudczanny–Passenheim, and further Rastenburg–Seeburg, whereas Samsonoff ordered the line Dombrowy–Mlawa as the objective for the 20th August.

Jilinsky had also instructed Samsonoff to push Cavalry, supported by Infantry, across the Frontier on the 18th. Samsonoff made no effort to carry out this order.

Here was a case in which some sort of Strategical Advanced Guard might have been employed. The enemy was some distance away, and the Advance to meet him was likely to be a long one. Many of the Corps were not ready, especially in their rear services. Samsonoff evidently intended to subsist on the country to a certain degree, and it was essential that a service of requisitioning should be organised in front of Corps. This could only be done behind a strong Cavalry screen. He made no attempt to send forward his Army Cavalry or to organise a requisitioning service. All Cavalry duties in front of his main advance were left to the weak and ill-trained Corps Cavalry. For his 4½ Corps this amounted to 22 Squadrons. It was also very unequally divided amongst the Corps (VIth Corps, 8 squadrons; XIIIth Corps, 5 squadrons; XVth Corps, 6 squadrons; XXIIIrd Corps, 1 squadron; and Ist Corps, 2 squadrons).

Samsonoff was attempting to issue orders to 9 units direct—6 Corps and 3 Cavalry Divisions,—strung out on a front of 135 miles as the crow flies. Jilinsky should have retained the IInd Corps under his own command, for Samsonoff's only means of communication with the Corps lay through the Headquarters of the North-West Front at Bialystok. As the Second Army advanced, the difficulties of communication with the IInd Corps would become greater.

The 4th Cavalry Division was directed into the most unsuitable country South of Johannisburg, where the Germans had erected a blockhouse line in the Forest. Against such an obstacle the Cavalry were useless. The country was so bad that it was not till two days after the Right Corps had crossed the Frontier that the 4th Cavalry Division succeeded in crossing at the same point. A special Infantry detachment from the Right Corps or from the IInd Corps, if it was under his command, would have been more suitable. The Cavalry would thus have been released for their legitimate work—reconnaissance in front of the Army.

Jilinsky had called Samsonoff's attention to the danger of attack against his left from the direction of Allenstein. Samsonoff was now directing two of his three Cavalry Divisions 60 miles to the West of Allenstein.

As in the case of Rennenkampf, Samsonoff was more anxious about the protection of his flanks than about the lack of information he had about the enemy.

Taking Samsonoff's direction as he ordered it, it is suggested that the Cavalry should have been grouped as follows :—

Right Group—

Strength : 4th Cavalry Division, closely followed by one Infantry Brigade (eight Battalions) from the VIth Corps as Advanced Guard to that Corps, supplied with a good proportion of the eight Squadrons of Corps Cavalry.

Objective : Dombrowy–Ortelsburg.

Information required—

(*a*) Did the enemy intend to oppose the Advance South of Ortelsburg ?

(*b*) What movement was taking place from East to West on the railway line through Ortelsburg ?

The railways leading out of Ortelsburg to be temporarily interrupted, and all rolling-stock in that place to be retained there.

Every facility to be given to the requisitioning officers moving in front of the VIth Corps. The utmost care to be taken to spare any supplies existing.

Left Group—

Strength: 6th Cavalry Division, closely followed by one Infantry Brigade (eight Battalions) from the XVth Corps as Advanced Guard to the Corps, supplied with a good proportion of the six Squadrons of Corps Cavalry.

Objective: Neidenburg.

Information required—

(*a*) Did the enemy intend to oppose the Advance South of Neidenburg ?

(*b*) What movement was taking place through Neidenburg on the railways ?

The railways leading out of Neidenburg to be temporarily interrupted, and all rolling-stock in that place to be retained there.

Every facility to be given to the requisitioning officers working in front of the XVth Corps. The utmost care to be taken to spare any supplies existing.

The 15th Cavalry Division should have been placed under the Commander of the Ist Corps, with Special Instructions as to the protection of the Left Flank of the Advance. It is suggested that he might have been issued an Instruction on the following lines :—

" The 15th Cavalry Division is placed under your orders. The Ist Corps will be responsible for the protection of the Left Flank of the Advance. The Corps will direct its march to the area of Soldau. Cavalry reconnaissances, backed by strong Infantry

Advanced Guards, will be directed towards Osterode, Deutsch Eylau, and Strasburg. The railways leading to Soldau from these three directions will be interrupted as far from Soldau as possible. It is important to ascertain—

(*a*) Whether forces are being directed from the Vistula towards Osterode, Deutsch Eylau, and Strasburg ?

(*b*) Whether there is a general movement, either by road or rail, from East to West through these three places ? "

The IInd Corps had been definitely tied to the Advance against the Masurian Lakes by Jilinsky. He gives his reasons for this Advance as the unfinished state of the fortress of Grodno. He undoubtedly feared some such stroke as Von Schlieffen had studied in 1901. He had falsely interpreted the meaning of such a German project. At the date upon which he issued his Instruction (13th August), the time for a stroke against Grodno had long gone past. The decisive battle in the West was expected about the 14th to the 17th August, and the Germans were unlikely to have detached troops which could arrive in the West in time for this battle. Strong raiding parties might well be expected, but the concentric advance of the two Russian Armies at either end of the Line of Lakes must necessarily call a halt in any large German offensive from the Lakes. The IInd Corps was a false detachment. The mission given to it could have been performed by a smaller force than a Corps of 32 Battalions and 14 Squadrons. Had it been Jilinsky's intention to deal with Fort Boyen and so force the Corps through the Lötzen Gap there might have been some justification for the detachment. But the attack of Fort Boyen was expressly excluded until further orders.

It is considered that a force of one Infantry Brigade (eight Battalions), with six Squadrons of Cavalry, would have been amply sufficient for the purpose of dealing with any German raids from the Lakes. A further Infantry Brigade might have been employed to clear the Johannisburg Forest, and the remainder of the Corps held ready to move as required by the course of events.

The Corps should certainly not have been added to Samsonoff's already unwieldy Army.

3. The Advance.

Samsonoff's Instruction No. 1 produced an immediate protest from Jilinsky, who wired on the night of the 16th :—

"Thanks to your extension to the West, you have increased your frontage at the passage of the Frontier to 40 miles, which I consider excessive. In view of the fact that the I[st] Corps is in Reserve to the Headquarters of the North-West Front, and you regard it as in support of the Second Army, I consider that you should move up the 2nd Division (XXIIIrd Corps) into the front line."

Samsonoff replied to Jilinsky that he was gaining more freedom of manœuvre by moving off to the

West, and thereby enveloping the enemy to a deeper extent.

On the 19th, the Commander of the North-West Front telegraphed to Samsonoff :—

"The delay in the advance of the Second Army is putting the First Army in a difficult position, and they have already been in action for two days at Stallupönen. Therefore hurry up the advance of the Second Army, and hasten your operations as energetically as possible, employing the Ist Corps if necessary."

Jilinsky was therefore urging the Second Army forward to its original timing, as he now saw that it was going to be late in arriving at the Frontier. He released the Ist Corps, thus giving Samsonoff complete control of the $4\frac{1}{2}$ Corps (VIth, XIIIth, XVth, Ist, and half the XXIIIrd). The IInd Corps was still tied to the Lakes on the right.

Samsonoff replied to this telegram :—

"The Army is advancing according to the time-table in your orders, without halting, covering marches of more than 12 miles over the sand; therefore I cannot go more quickly."

No contact had been established with any enemy, and nothing was known about his grouping. By the night of the 19th/20th the Cavalry on the left flank reported that the enemy appeared to be concentrating in the Neidenburg–Soldau–Gilgenburg area, otherwise nothing.

By the evening of the 20th the Army was situated as follows :—

IInd Corps—Jucha–Rosinsko.
4th Cavalry Division—Kolno.
VIth Corps—Dombrowy–Pelty.
XIIIth Corps—Zaremby–Chorzele.
XVth Corps—Bajki–Zaboklik.
2nd Division (XXIIIrd Corps)—Murawy.
6th Cavalry Division—Kuklin.
Ist Corps—Tezcianko.
15th Cavalry Division—Nowawies.

The right had reached the Frontier, but the left was still some distance from it. The 2nd Division (XXIIIrd Corps) was well behind the main line.

For the 21st the Army orders were as follows :—

"The VIth, XIIIth, XVth, and Ist Corps will reach the line Friedrichsfelde - Wallen - Kutzburg - Kannwiesen - Janowo - Nowawies - Mlawa. The XXIIIrd Corps (2nd Division) will move into the Koziczyn area; the Guard Corps to the Mlotsk area" (near Warsaw).

Samsonoff reported to the North-West Front on the night of the 20th :—

"The frontier roads are very heavy. Reconnaissances therefore were only short. Enemy aeroplanes followed the movements of the Army all the time."

On the 21st, Jilinsky withdrew the IInd Corps from the Second Army, and transferred it to Rennenkampf. He also informed Samsonoff that the Guard Corps was not under his command, as he appeared to think. He gave him the 1st Rifle

Brigade. Samsonoff now had his $4\frac{1}{2}$ Corps; the 3rd Guard Division (the other Division of the XXIIIrd Corps) and the 1st Rifle Brigade being *en route* to Novo Georgievsk, and not for the moment available.

By the night of the 21st the VIth and XIIIth Corps were well across the Frontier, with the 4th Cavalry Division at Demby. No contact had been obtained with the enemy. The XVth and Ist Corps were now close to the Frontier, with the 2nd Division (XXIIIrd Corps) well behind at Koziczyn. There were indications of enemy troops at Neidenburg.

For the 22nd the Army Orders were as follows :—

> Ortelsburg–Kutzburg–Kannwiesen–Neidenburg–Soldau.
> The XXIIIrd Corps (2nd Division) to Mlawa.
> The 4th Cavalry Division to Lipniki–Demby.
> The 6th Cavalry Division to support the attack on Soldau.

This order at once drew the following telegram from Jilinsky :—

" I consider the dispositions given by you for the 22nd most undecided. I must have immediate and decisive operations."

General Samsonoff reported that the reasons for delay were as follows :—

" The great weariness of the troops, the necessity for bringing up the XXIIIrd Corps (2nd Division), the unwieldiness of the rear services, and the general incompleteness of units."

The real truth was that the Army was now in a terrible state. Since coming into line the marches had not exceeded 9 to 12 miles, but all units had been marching through a sandy desert in a country that could not support them, and without any food being brought up from the rear. Some units had been marching ten to twelve hours a day for eleven days, and all had marched from the 15th to the 22nd without a halt.

Observers reported conditions as follows :—

"I arrived at General Martos' Headquarters (XVth Corps) at 8 A.M. The Corps had begun its movement on Neidenburg in four columns. The movement was very halting, order bad. Many stragglers. All were asking for a halt."

Samsonoff reported :—

"It is essential to organise the rear services. This has up to the present not been done. The country is devastated. The horses have long been without oats, and there is no bread. It is impossible to bring them up from Ostrolenka."

General Klueff (XIIIth Corps) reports :—

"The Corps transport was disembarked far in rear, and could not be brought up. During an uninterrupted forward movement beginning with Ostrolenka, on unmade roads, it was impossible to organise traffic properly, despite every effort.

"Little came through, and that only by chance. Of local resources cattle only, with some poultry. Hay and oats not yet cut. All the villages were deserted. In eleven days we have covered 132 miles over heavy sandy roads."

A General Staff Officer reports :—

"The Infantry and Artillery were miserably provided for. I don't know how the troops bear it any longer. It is essential to organise a proper requisition service."

The XXIIIrd Corps reported :—

"Owing to the taking away of my rear services, the provisioning of my Corps has never yet taken place."

Staff work, never good, was reported as deteriorating, and much quarrelling was going on.

A General Staff Officer of the Army reported :—

"I saw the work of the XVth Corps Staff. The work is heavy, and the Corps Commander (Martos) very quarrelsome and incredibly rude. Everything is at loggerheads."

And again :—

"The left flank Corps Commanders are certainly the most incompetent (Artamonoff and Kondratovitch)."

The advance was just at breaking-point, without having been in touch with the enemy. The Higher Commanders, oblivious to the state of the Army, were too proud to acknowledge the condition of the men and call a halt, cost what it might.

The Army, even on the 22nd, was unfit to fight.

It may be an apt moment to recall what is termed the waning power of the offensive. As an advance continues, more and more men are required for the guarding of strategic points in an enemy's country, for the direction of traffic, and for the patrolling of roads. Casualties and sickness reduce numbers in an astonishing proportion. Commanders often regard their units as at full strength continually, and proceed to allot them tasks above their strength to perform. The less organised an Army is as regards its rear services, the greater is this wastage in an advance.

It therefore often comes to a point when an advance must automatically cease of its own accord.

It is interesting to compare the numbers of men which Napoleon led across the Niemen in 1812, and those which reached Moscow. In modern times we have the very definite examples of Kolchak advancing from Siberia or the Soviet troops advancing against the Poles in Warsaw. If the defender can time his blow so that it takes place at the limit of extension of the advance, the results of his counter-attack are almost invariably annihilating. There can be no doubt that the German advance in 1914 on Paris and beyond had almost reached its limit, and it was only due to the excellent organisation of the German communications that they were enabled to recover from Gallieni's blow from Paris and the Battle of the Marne.

The efficient rear organisation of the First German Army during its march through France and Belgium, as given by Von Kluck, should be noted. On the 22nd August the Supreme Command handed over 20 Battalions of Landsturm and 4 Squadrons of Landsturm Mounted troops for L. of C. work. All the Landwehr thus freed from this work were sent to reinforce the IIIrd Reserve Corps, which was covering the rear of the First German Army. Light railways were opened up for each of the three Corps, and 38 motor columns and 15 Reserve Supply Columns were brought into action.

A comparison of such an organisation, working on good roads, with the organisation possessed by the Russian Army, working on sandy tracks, makes clear the task which had been set to the Second Russian Army.

The fighting on the 22nd was limited to a skirmish at Soldau, which was occupied by the Ist Corps.

Orders for the 23rd were as follows :—

VIth Corps—Ortelsburg.
XIIIth Corps—Jedwabno.
XVth Corps—Neidenburg.
XXIIIrd Corps (2nd Division)—Koslau.
Ist Corps—Soldau.

It will now be seen that Samsonoff had only one Corps on the line of his original objective. All the remainder were well to the West of it. He had deliberately disobeyed his orders, and Jilinsky, instead of ordering, had merely protested and given him another Corps. Samsonoff, like Jilinsky, was becoming more grandiose in his ideas of a turning movement. Each was neglecting the administrative difficulties arising from a change of direction.

As the General Staff at Front Headquarters expressed it, " Samsonoff was pulling the Army to the left, and Jilinsky was pulling it to the right."

The enemy was being given more and more elbow room between the two Russian Armies.

The objectives of the Second Army were reached

on the night of the 23rd, and on this night Rennenkampf's First Army occupied the line Peleningken–Darkehmen.

Information about the enemy was accumulating. Two Landwehr Brigades were moving from Soldau to Gilgenburg. Two bivouacs, each of a Division, were near Gilgenburg. On the right flank no information. The 4th Cavalry Division had just reached Ortelsburg, and had made no reconnaissance forward. Everything seemed to point to the fact that the Germans had slipped past to the West, and that there was little likelihood of enveloping them and shutting them in against the Lakes.

On the 23rd, therefore, General Samsonoff, through his Quartermaster-General, General Filimonoff, asked the concurrence of the Commander of the North-West Front in a deviation from his Instructions to advance to the line Rastenburg–Seeburg in favour of an advance to the line Allenstein–Osterode, for the following reasons :—

1. The Army could better fulfil its task of getting astride the rear of the troops defeated by the First Russian Army at Gumbinnen.
2. The Army could be better based upon the Soldau–Mlawa Railway.
3. It would be easier to advance into the heart of Germany from the line Allenstein–Osterode than from the line Rastenburg–Seeburg.

In actual fact, Samsonoff had already committed himself, and had made no attempt to advance in the direction he had been ordered to take. Jilinsky knew this well, and had already protested against the Westward tendency of the advance.

Jilinsky did not agree to Samsonoff's demand, and censured him on the telephone through the Quartermaster-General.

But during the 23rd Jilinsky received encouraging news from Rennenkampf, and wired to Samsonoff:—

"German troops, after heavy fighting and the victory of General Rennenkampf, are hastily retreating, blowing up the bridges behind them. The enemy apparently has only inconsiderable forces in front of you.

"Therefore, leaving the Ist Corps at Soldau and protecting the left flank with the necessary troops, you will advance energetically with all remaining forces to the front Sensburg–Allenstein, which should be occupied not later than the 26th.

"Your movement is directed to meet the enemy retreating in front of General Rennenkampf, and to cut off his retreat to the Vistula."

Jilinsky had become infected with the optimism of Rennenkampf, who, on the 23rd, had recommenced his advance without gaining touch with the enemy, but finding every sign of a hasty withdrawal.

Jilinsky had given in to the demand of Samsonoff, ordering him to turn slightly more Westward. Why he did not order Samsonoff to conform at once to this new direction, from which he knew he had already deviated, is difficult to understand.

A further complication arose from the fact that by some means or other Sensburg and Seeburg were confused in the telegraphic orders. Considerable telegraphing on this matter took place between the

North-West Front and the Second Army. The North-West Front eventually stated that Sensburg was intended and not Seeburg. Samsonoff very properly pointed out that Jilinsky was now indicating a frontage for the Army of 35 miles, whereas he had already been complaining of such an extension of front. The I[st] Corps was to be isolated in Soldau, while the advance was to continue with 3½ Corps (VIth, XIIIth, XVth, and half the XXIIIrd).

The original Army had consisted of 6 Corps and a Rifle Brigade, so that from one cause and another the final advance of the Army was to be carried out by a force reduced to half its original size.

Jilinsky's wire contains also the first definite date after crossing the Frontier to which the Army had to conform. Samsonoff was to reach Sensburg with his right flank on the 25th. Rennenkampf might reach Angerburg on that date, when the flanks of the two Armies would be 29 miles apart.

Samsonoff, however, still stuck to his idea of gaining ground to the West, and made no attempt to carry out Jilinsky's orders.

On the 24th the Army was to occupy the line Ribben-Gieslingen-Scheufelsdorf-Layss-Kellaren-Ganglau-Waplitz. On the 25th, Sorquitten-Bischofsburg-Wartenburg-Allenstein-Hohenstein.

Still only the Corps Cavalry in front of the Army, and the Army Cavalry wide to a flank.

The 4th Cavalry Division was ordered to Sensburg on the 24th, with reconnaissances to the North between the lines Rastenburg-Gerdauen and Rössel-Schippenbeil. Observation was also to be made towards Lötzen. Again a double objective.

The 6th and 15th Cavalry Divisions were to reconnoitre to the line Heilsburg–Zinten (S. of Königsberg), cutting communications to the Vistula.

The order for the 24th and 25th concluded by saying :—

"When these operations have been concluded, the troops will maintain themselves preferably from local resources."

This is what they had been doing, or rather not doing, for ten days.

The advance was being continued without any knowledge of what had become of the enemy who had left the field of Gumbinnen. The reconnaissances ordered for the 4th Cavalry Division to the North were not carried out, and most of the energy of this force was expended in watching Lötzen. Here again no detailed instructions to the Cavalry were issued. Had they been given definite questions to answer, they might have carried out their work more efficiently. The main thing that Samsonoff required to know was at what point the Germans had struck Westwards. It is suggested that the 4th Cavalry Division should have been ordered to reach the line Rastenburg–Korschen, destroying the railways, and to find out if the Germans were still in force to the East of this line, or whether they had already passed to the West.

The frontage of the Army was now 60 miles, that of Corps 10 to 13 miles.

On this subject Jilinsky wired to Samsonoff on the 24th :—

"Earlier I directed your attention to, and now

do not approve of, the extended condition of your front. For further operations, close your Corps to more narrow frontages, so that they do not have to fight isolated, and are not out of liaison with one another."

4. The Battle of Orlau–Frankenau.
(Sketch D.)

On the 24th the XVth Russian Corps attacked the Germans on the line Orlau–Michalken. The remaining Corps of the Army reached their objectives without incident.

The fighting on this day is another example of what occurs when there has been no reconnaissance and the troops are too extended.

The German position ran astride the main Neidenburg–Allenstein road, some 12 miles in length. A line of low hills, 60 to 100 feet above the plain, runs from Orlau *viâ* Lahna and Frankenau to Michalken, where they abut closely upon a small lake. The field of view and fire over the plain towards Neidenburg is excellent. The plain is covered with potato-fields, offering little or no cover, the sandy fields making any movement plainly visible through the dust raised. Only opposite Orlau can the position be approached under cover, and there the German position held the exits from the Kommusin Forest under effective rifle-fire.

Every movement of the XVth Russian Corps (8th and 6th Divisions), as it emerged from Neidenburg on the morning of the 24th, must have been plainly visible. The Russians made no attempt to locate the enemy, and their right flank (8th Division) appears to have blundered up against very strong positions at Orlau, suffering heavily as they left the woods, and finally being brought to a standstill by a German counter-attack with the bayonet. Fighting proceeded here during the whole of the 24th without dislodging the Germans.

There are nearly 1,000 Russians buried in the cemetery at Orlau, and only some 200 Germans.

Lahna, which is the highest point of the line, was equally unassailable, and the Russians lost heavily here also.

On the left the 6th Russian Division attacked Frankenau and Michalken, inflicting casualties upon the Germans, but failing to drive them out of their position.

The Commander of the XVth Corps appealed for help to the 2nd Division on his left, then at Roggenhausen. This Division moved up to Lippau, with its advanced guard at Skottau, with a view to attacking the German right at Michalken, when an urgent message arrived from the Ist Corps, still farther to the left near Usdau, asking for help also. The 2nd Division turned to the West, and commenced to take up a line between Kownatken and Thalheim. A sudden cry of " Cavalry " created a panic, and the Division commenced to break. Indiscriminate firing followed, in which there were 100 casualties, and the Artillery ammunition waggons dashed through the Infantry in reserve and dispersed them. There was no enemy in sight. Thus no help arrived from the 2nd Division on the 24th.

On the right of the XVth Corps the Commander of the XIIIth Corps, hearing the battle commencing on his left, decided to move down on the enemy's left flank. On the morning of the 25th his left Division reached Lindenwalde by 5 A.M., having driven in some of the Germans during the night. The other Division of the XIIIth Corps remained at Persing. But it was too late, and the Germans were found to have evacuated the position on the early morning of the 25th, making good their retreat towards the N.W.

The German forces were largely composed of Landwehr units, more especially at Michalken, and had carried out their delaying operations well. With inferior numbers they had inflicted heavy casualties on the Russians.

The XVth Russian Corps lost 50 officers and 2,500 men on this day, capturing two guns and a few prisoners.

Had the Russians made a proper reconnaissance, the march of the XIIIth Corps and the 2nd Division might have been so directed from the outset as to converge upon the flanks of the German position. As it was, they were too far away to co-operate in time.

On the 25th the following telephonic conversation took place between Samsonoff and the Chief of the Staff of the North-West Front:—

Samsonoff. "The enemy about three Divisions strong, after the battle with the XVth Corps and the outflanking of the XIIIth Corps, withdrew in the direction of Osterode. If I obey my Instructions, it is possible to leave a screen against him and to continue the advance on Allenstein and Sensburg, but then it will not be possible to pursue the retreating enemy. I therefore ask to be allowed to deviate from my Instructions, and that I may advance on the line Allenstein–Osterode."

The reply shortly afterwards from the Commander of the North-West Front was as follows :—

"If it is certain that the enemy is retreating on Osterode, the Front Commander agrees to the alteration in the advance of the Second Army to the direction of Allenstein–Osterode, but under conditions that the movement between the Lakes and Allenstein is covered by a Corps and a Cavalry Division, which should preferably be at Sensburg."

Once more Jilinsky yielded to the Westward pull of Samsonoff. Samsonoff had had no verification of his assumption that the enemy had retired to Osterode. Jilinsky evidently was not sure that Samsonoff was right, but he merely orders a Corps and a Cavalry Division to be isolated from the Army at Sensburg. It is difficult to conceive what this detachment was for. If the Germans in front of Rennenkampf were retreating hastily, they were not likely to move anywhere but towards either Königsberg or the Vistula. The number engaged at Gumbinnen must have been known with fair accuracy, and there could not now be a large force near Lötzen. If the Germans withdrawing from Gumbinnen were not in disorder, and were contemplating attacking the right flank of the Second Russian Army, then the proper place for the right flank defence was closer in to the remainder of the Army. Jilinsky had ordered Samsonoff to isolate a Corps from the others, a thing that he had been complaining Samsonoff had done before. The more meagre the information regarding the enemy's intentions, the more reason was there for holding the Second Army together.

Samsonoff ordered the VIth Corps to Bischofsburg and the 4th Cavalry Division to Sensburg, but there are no separate Instructions for this detachment. They are "to protect the right flank of the Army." All Commanders of detachments must be given definite instructions as to what they are to do. If they are attacked in superior force, they must be told how long they are to hold out, and whether they are to maintain their positions until reinforced. If they are forced to retire, they require to know the direction in which they are to go. The VIth Corps Commander received no such Instructions.

5. Position of the Second Russian Army before Tannenberg.

By the evening of the 25th the Second Army had reached the following positions:—

Right Flank.
VIth Corps—Bischofsburg.
4th Cavalry Division—Sensburg.

Centre.
XIIIth Corps—Kurken.
XVth Corps—Orlau-Frankenau.
2nd Division (XXIIIrd Corps)—Lippau.

Left Flank.
Ist Corps—Usdau–Meischlitz–Koschlau–Grallau.
6th Cavalry Division—Gr. Lensk.
15th Cavalry Division—Zielun.

The 3rd Guard Division of the XXIIIrd Corps had now one Regiment—the Litovski Guard Regiment—at Koslau, with the remainder at Mlawa and Warsaw. The 1st Rifle Brigade was at Novo Georgievsk.

Information about the enemy may be summarised as follows :—

1. Two Brigades with Artillery had left Allenstein in a Westerly direction on the morning of the 25th.
2. A bivouac of about one Division lay between Gardienen and Lake Gross (E. of Gilgenburg).
3. Forces of all arms were moving from Strasburg on Lautenburg.
4. Landwehr and Reserve troops in large numbers at Neumark and Strasburg.
5. Large forces moved through Rastenburg in a Westerly direction on the 24th.

On the morning of the 26th Samsonoff issued Order No. 6353, as follows :—

" In front of the XVth Corps and 2nd Division, no change. At Allenstein about one enemy Division.

The Second Army will continue its advance on Allenstein–Osterode.

XIIIth Corps—Kellaren.
XVth Corps—Schönfelde–Gusenofen.
2nd Division (XXIIIrd Corps)—Hohenstein–Reichenau.
Ist Corps—Soldau–Gilgenburg.
VIth Corps—Bischofsburg."

Apparently after the issue of this order, the news of the 25th, that the Germans were concentrating in the direction of Lautenburg, was considered, and Samsonoff decided to hold up the Advance. Unfortunately he was overpersuaded by his Chief of the Operation Section, Colonel Viyaloff, and finally decided to reinforce his left. The Ist Corps Commander was therefore given command of the remainder of the 3rd Guard Division, the 1st Rifle Brigade, and the 6th and 15th Cavalry Divisions, with some additional Heavy Artillery. This force might well have been considered sufficient, but the possibility of concentrating it only existed on paper. The Heavy Artillery, 1st Rifle Brigade, and the 3rd Guard Division were not within tactical supporting distance.

The picture is now complete.

Samsonoff, in complete ignorance of the German stroke descending upon him, was about to continue his advance against the line Allenstein–Osterode with 3½ Corps and 2 Cavalry Divisions, leaving 1 Corps and 1 Cavalry Division isolated on his right at a distance of two days' march.

His troops were tired and unfed. Stragglers were numerous, and the Army much reduced by sickness. Only the XVth Corps had been engaged with the enemy, losing some 2,500 men. All had been marching over execrable roads since the 15th, and some even before that date.

6. Samsonoff's Signal Communications.

One of the many causes of the failure of Samsonoff's operation was a lack of good means of intercommunication in the field. He had no direct communication with Rennenkampf, and Jilinsky failed to keep him informed of the movements of the First Army. But in the Second Army itself communication was deplorable.

The signal material in the Army consisted of the following units :—

(i) An Independent Telegraph Company.
(ii) A Column for repairing telegraph lines.
(iii) An establishment of 25 telephone instruments with the necessary operators.
(iv) Two field Radio stations.

The operators were, generally speaking, very inexperienced, and only about 80 miles of cable was available.

The mainstay of the communication system was the existing net of Government telegraph lines. These ran as follows :—

(i) Ostrowo–Ostrolenka–Myszyniec.
(ii) Ostrowo–Makow–Przasnysz–Mlawa.
(iii) Przasnysz–Chorzele.

The above were the only stations existing in the area of the advance.

The Army did not consider itself responsible for the arrangement of communication with Corps, which were required to continue the lines from the final Government stations with their own 50 miles of cable. Nothing was therefore left for communication with their Divisions.

Even before the 20th there was much delay in messages, but after that date the whole system broke down badly.

The VIth Corps worked to Myszyniec, the XIIIth and XVth Corps to Chorzele, while the Ist Corps used Mlawa.

The Army Independent Telegraph Company was employed to lay a line from Ostrolenka to Przasnysz and thence to Mlawa, with a view to supplying lateral communication. By the time this work had been completed and the Company was at Mlawa, an order was received to take down the line and proceed to Ortelsburg, where Army Headquarters was supposed to be going. On its way to Ortelsburg, at Chorzele, an order was received to go to Neidenburg. The Company eventually arrived at Neidenburg after Samsonoff had left and the fate of the Second Army had been decided.

When Samsonoff moved up from Ostrolenka to Neidenburg, in order to be able to direct his Corps to better advantage, he took with him what he termed his Field Headquarters only. He then found himself practically cut off from his administrative officers. Messages from Neidenburg to Ostrolenka had to be transmitted through five stations—Mlawa–Przasnysz–Makow–Ostrow to Ostrolenka.

The VIth Corps, having used up all their cable between Myszyniec and Ortelsburg, had to rely on mounted orderlies from Ortelsburg forward.

One has become so used to an efficient system of signals that the state of communications in the

Russian Army can hardly be realised by a British officer.

The order giving the dispositions of the Second Army for the 24th was sent out in clear, and it is therefore not surprising that the Germans were able to make dispositions to envelop the wretched Russian Army.

CHAPTER VI.

The German Advance against Samsonoff: Battle of Tannenberg.

Map II.
Sketches E, F, G, H.

1. The Situation on the 22nd August.

When Ludendorff reported at Coblentz, in accordance with orders appointing him Chief of the Staff in East Prussia, he found that the situation in the East was regarded as extremely serious. In notifying him of his appointment, General von Stein had used the words, "You may yet be able to save the situation."

General Von Prittwitz had been superseded, and the Eighth German Army was thought to be withdrawing rapidly behind the Vistula. The name of the new Army Commander was not yet known.

Supreme Headquarters knew nothing, as we know, of any change in the situation since the first panic report.

Ludendorff at once began to issue orders. He evidently considered that the Headquarters of the Eighth German Army were out of the picture, for he ignored them entirely, issuing orders direct to units in the Army. He tells us that the chance offered to the strategist was one not to be missed. The Second Russian Army was for the moment the most menacing, and must be dealt with first.

His orders were as follows :—

(i) Army Headquarters to meet him at Marienburg.
(ii) The XVIIth and Ist Reserve Corps to halt during the 23rd.
(iii) The Ist Corps to detrain at Deutsch Eylau and South of it, and commence moving up on the right flank of the XXth Corps.
(iv) The 3rd Reserve Division to detrain at Allenstein, and move up on the left of the XXth Corps.
(v) A Landwehr Division from Schleswig-Holstein to entrain for the East.
(vi) All available garrisons from the Vistula to move up on the right of the Ist Corps, to strengthen the garrison troops at Strasburg and Lautenburg.

There was no mention of other reinforcements at the moment.

It is interesting that Ludendorff and Hoffmann had issued practically the same orders. The credit of instantly realising the situation cannot, however, be taken from Ludendorff. He assumed the responsibility for issuing orders without reference to the new Commander, whoever he might be. That these two men came to the same conclusions to meet the situation is only a proof of the high training existing in the German General Staff.

Hoffmann had ordered the Ist Corps to detrain

at Gosslershausen, slightly farther West than Deutsch Eylau, and the Army Headquarters moved to Muhlhausen, slightly farther East than Marienburg. The 3rd Reserve Division had already been ordered to the left of the XXth Corps, instead of to its right, upon the request of the Commander of that Corps.

Hoffmann's main idea was to check the advance of Samsonoff's left, in order to disengage from Rennenkampf. He tells us that his idea was to concentrate the Eighth German Army in the region of Osterode, from whence a blow might be dealt later against either Samsonoff or Rennenkampf.

Ludendorff's idea was to deal an immediate blow against Samsonoff.

The movements ordered by Hoffmann undoubtedly facilitated the movements ordered by Ludendorff, for little change was necessary.

Hindenburg and Ludendorff met for the first time on the railway station at Hanover on the evening of the 22nd. It is reported that Hindenburg had only been fitted that very day with his new "field grey" uniform. At 2 P.M. on the 23rd they reached Marienburg and assumed control.

Ludendorff tells us that he met with a frigid reception.

He makes no mention of Hoffmann's changes in the original order to withdraw to the Vistula, and yet these changes must have been made, for a study of the railway orders reveals the fact quite plainly. Ludendorff was never inclined to give credit to others where he himself might be belittled.

Operations began with energy.

On the evening of the 24th the Army telegram

from Marienburg to Supreme Headquarters shows the plan taking shape :—

"Have decided to hold fast with the XXth Corps, as withdrawal will have bad moral effect. Transportation of Ist Corps being delayed. The Ist Reserve and XVIIth Corps will be drawn towards the left flank of the XXth Corps. Good spirits here if worse befall."

The XXth Corps was to form the framework for the fresh operations. Upon each of its flanks were to be moved strong forces, so directed from the outset that they would take in flank the forces advancing against this Corps. Rennenkampf was to be opposed by a skeleton Cavalry force.

The whole success of the operations depended upon whether Rennenkampf would allow himself to be deceived and would fail to hurry to Samsonoff's assistance.

The operations of the XXth German Corps had been so far as follows. Originally in position near Ortelsburg, the Commander had given up the idea of attacking Samsonoff in the early days of his advance, owing to the extent of the front and the Russian numerical superiority. He had withdrawn to the line Frankenau–Orlau, where he had fought a battle on the 24th. He had then withdrawn to the line Gilgenburg–Mühlen–Hohenstein.

There had never been any question of this withdrawal having been according to a preconceived plan to lead the Russians into a trap laid by Hindenburg in time of peace, as has been asserted by both German and Russian writers. It was an unwilling retirement in the face of great superiority. The XXth Corps had been hard-pressed, but had delayed the enemy considerably.

On the 24th the Russian wireless gave the Germans complete information as to their proposed movements on the 24th and 25th.

2. The Disengagement after Gumbinnen.

The disengagement of the German forces after Gumbinnen had been cleverly effected, helped by the passive attitude of Rennenkampf, who made no move forward until the late morning of the 23rd.

The I^st Corps on the left, having more or less disorganised forces opposing it, was the first to go. Von François placed Von Brodrück's Königsberg Garrison in position East of Insterburg to cover entrainment, and marched his two Divisions Westwards. The front was taken over by the 1st Cavalry Division.

Entrainment took place at Insterburg, Norkitten, Wehlau, Tapiau, and Königsberg during the period from the 21st to the 23rd. The work was delayed by the fact that some of the station detachments had been too hurriedly withdrawn to Königsberg in the prevailing panic, and these stations could not be used for entraining.

It was during this period of the 21st to the 23rd that the German Supreme Command, having superseded Von Prittwitz, entered into direct communication with the Commander of the I^st Corps, asking him what he thought of attacking Rennenkampf once more. Von François replied that the moment had gone by, and that there was nothing left but to concentrate against Samsonoff.

Masses of fugitives were now crowding the railways, and rolling-stock for the Ist Corps had to be brought from West of the Vistula, still further delaying the movement.

Detraining stations were Neumark, Deutsch Eylau, Osterode, and Biessellen. Detrainment was completed by the evening of the 26th, though all but certain Artillery units were off-loaded by the 25th.

The route followed was Königsberg–Marienburg.

The capacity of the East Prussian railways is shown by the fact that, under great stress, they were able to switch the Ist Corps, firstly from the Vistula to Gosslershausen, and then to Deutsch Eylau and South of it. All this during actual train movement.

The 3rd Reserve Division on the right was the next to go. This Division stood ready to attack on the morning of the 21st near Benkheim. It marched the 13 miles to Angerburg, and 20 to Nordenburg, and commenced entraining at once. Vehicles entrained at Lötzen and Rastenburg. The route followed was Korschen–Allenstein. The whole Division was detrained by the night of the 23rd, though the fighting units were assembled at Allenstein by the night of the 22nd. The entrainment had taken place within 30 miles of an enemy who had not been engaged on the 20th at the Battle of Gumbinnen. The distance from Angerburg to Allenstein is 79 miles by rail, and the time employed to transport the complete Division (48 hours) is a good example of the time it takes to carry out an operation of this nature. The actual time taken on the journey is small in comparison with the time employed in entraining and detraining.

When it is remembered that the Russians had no railways available for such movements, it is clear at what disadvantage they were.

The train movement of Von der Goltz's Landwehr Division is also interesting. The entrainment was at Kiel–Neumunster, and the route followed Hamburg–Wittenburg – Berlin – Schneidemuhl – Marienburg–Biessellen.

Entraining commenced on the 22nd, and the movement was complete by the 28th. The distance in this case is 550 miles.

The Ist Corps and the 3rd Reserve Division were at once pushed up on either side of the XXth Corps, the 3rd Reserve Division on the left and the Ist Corps on the right.

The XVIIth Corps was disengaged next, passing behind the Ist Reserve Corps, the 1st Cavalry Division extending Southwards as the XVIIth Corps left the line. The Corps reached the region of Muldszen, 37 miles from the battlefield, on the night of the 22nd, showing the energy employed to disengage. Friedland was reached on the night of the 23rd. By the evening of the 25th the XVIIth Corps had reached the region of Gr. Schwansfeld. It had covered 100 miles in five days in burning August weather.

The Ist Reserve Corps was the last to go. It reached Nordenburg on the night of the 22nd, 25 miles from the battlefield.

There was now nothing in front of the Russians but the 1st Cavalry Division.

The I[st] Reserve Corps then passed in front of the XVIIth Corps, somewhat delaying the movement. This crossing of the two Corps is a curious manœuvre. The withdrawal from the field of Gumbinnen in succession from the left was clearly indicated, but unless it was desired to have the Regular Corps on the left of the new operations, the movement is inexplicable. The XVIIth Corps had not behaved well at Gumbinnen, though it was probably superior in fighting value to the Reserve Corps.

3. The Attack on the Russian Right (Sketch E).

Orders to the 35th Division of the XVIIth Corps, issued at Gr. Schwansfeld at 10.50 P.M. on the night of the 25th, ran as follows:—

"The advance of three enemy Corps against the XXth Corps has been defeated. At Bischofsburg is situated a Russian Corps, the advanced troops of which are at Gr. Bessau. At Sensburg, enemy Cavalry.

The XVIIth Corps will advance to-morrow to Bischofsburg, with the object of attacking this Russian Corps, in co-operation with the I[st] Reserve Corps and the 6th Landwehr Brigade, now at Lautern.

The Corps will advance upon one road, the 36th Division leading, followed by the 35th Division."

Order No. 6346, of the Second Russian Army to

the VIth Corps, is dated 3.30 A.M. on the 25th. It was apparently despatched at 8.30 A.M., and reached the Corps during the afternoon.

"The VIth Corps, under whose orders is placed the 4th Cavalry Division, will occupy the area Bischofsburg–Rothfliess, with the object of protecting the right flank of the Second Army from Rastenburg. The 4th Cavalry Division will remain at Sensburg, reconnoitring between the lines Rastenburg–Bartenstein and Seeburg–Heilsburg."

On the morning of the 26th August the VIth Corps was disposed as follows:—

16th Division—Bischofsburg.
Advanced Guard—Adamshof.
4th Division—Rothfliess.
Advanced Guard—Kleisack.
General Line of Outposts—
Allmoyen–Striewo–Teistimmen.
Corps Cavalry—Bansen.

No reconnaissance had been carried out to the North; the Chief of the Staff of the 4th Division, Colonel Serbinovitch, pathetically observing, "No orders for reconnaissance in this direction had been received."

At some period on the 25th, after Samsonoff had received news of the XVth Corps fighting at Orlau–Frankenau, he telegraphed to the VIth Corps as follows:—

"Leave a screen to cover operations at Bischofsburg, and move with all speed to Allenstein to support the advance of the other Corps of the Army."

This order was received late on the 25th. It was, however, countermanded by another order, issued on the 26th. This counter-order reiterated the original order to remain at Bischofsburg for the protection of the right flank of the Army. It was only received after the battle of 26th was over.

The VIth Corps Commander, in accordance with his orders, was preparing to march the 16th Division, together with a Brigade of the 4th Division, South of Lake Dadey *viâ* Martinsdorf and Kirschbaum to Allenstein. Early on the morning of the 26th the unexpected news arrived that an enemy Brigade was at Lautern, and later that the enemy was occupying Seeburg. Lt.-General Komaroff, Commander of the 4th Division, thinking that this was a screen thrown out by the enemy retreating from Rastenburg in front of Rennenkampf, decided to attack. He moved his Division to the outpost line, disposing it on either side of the village of Teistimmen, with a detachment to protect his rear towards Seeburg.

Fighting at once became general.

The Corps Commander, in view of the heavy fighting with the 4th Division, cancelled the order for the move of one of its Brigades, but allowed the 16th Division to start upon its march to Allenstein about noon.

As the 4th Division became more and more seriously engaged, the Corps Commander realised his mistake and ordered the 16th Division to return to Bischofsburg. The order arrived when the Division had covered eight miles on its way to Allenstein. On its return, within four miles of Bischofsburg, it was met with a further order, to leave one Regiment and two Batteries in Corps Reserve and to move Northwards, West of Lake Dadey. Just as it was getting dark, near Ramsau, the Division ran into an enemy force of Infantry and Cavalry and came to a standstill. It had covered 19 miles since noon on that day.

The 4th Division was being attacked by the 36th German Division and the 6th Landwehr Brigade advancing from Lautern on Teistimmen, whilst the 35th German Division was advancing against their left flank and rear at Sauerbaum. The Ist Reserve Corps was that which had met the 16th Division at Ramsau.

The 4th Division had suffered heavily, and hearing nothing of the 16th Division, Komaroff ordered a withdrawal on Bischofsburg at 6 P.M. on his own initiative. There was a good deal of confusion, and the order did not reach the Reserve till 10 P.M. The Reserve reached Bischofsburg at 3 A.M. on the 27th, only to find that the Division had been ordered on to Sczepanken.

Finding that one of his Divisions had begun to disintegrate, the Corps Commander seems to have lost his head. Two orders were sent to the 16th Division during the night of the 26th/27th—first to concentrate at Martinsdorf, and second to concentrate at Theerwischolla. He did not notify the Army of the withdrawal of his Corps until 2 A.M. on the 27th, Samsonoff receiving the news about 12 noon on the same day. His telegram is sufficiently indicative of lost *moral* :—

"I have lost touch with the IInd and XIIIth

Corps. Troops became mixed up with the transport during the night, the latter having been brought up too close to Bischofsburg."

On the 26th the 4th Division lost 73 officers, 5,283 men, 16 guns, and 18 machine-guns.

As a result of the fighting on the 26th, the VIth Corps had become so demoralised that with little threat from the enemy it continued its retreat to Olschienen and Wallen. It is interesting to note that the demoralised Corps Commander called a Council of War consisting of his Brigadiers from each Division on this day. This was not the moment for taking advice from subordinates.

During the retreat on the 27th the troops became inextricably mixed, and on one occasion the arrival of the Corps Commander's motor-car, escorted by its Sotnia of Cossacks, raised a cry of "Cavalry." In the resulting panic there were considerable casualties, and the incident must have still further demoralised the Corps Commander when he saw the condition to which his troops had been reduced.

The story of this disaster to the Russian right is a melancholy one. Communication with the Army was bad, and Samsonoff issued order and counter-order without knowing whether such orders had reached their destination. All the Commanders were under the impression that they were dealing with a defeated enemy, who was retreating across their front. Neither tactical nor strategical reconnaissance had been made in front of the Corps, with the result that the 4th Division was surprised by a superior enemy, whilst the 16th Division was marching and counter-marching beyond supporting distance. A

Division withdrew without orders from the Corps, and the Corps Commander acquiesced in the retreat of the whole of his Corps in disorder to a distance as the crow flies of 21 miles. All touch with the Army was lost, and timely information of the disaster was not sent to his superior.

The incompetence of the Commander of the VIth Corps and of one of his Divisional Commanders was such that it may be argued that they would have failed, no matter what their orders had been. Nevertheless, it is in Order No. 6346 of the Second Army that the germ of the disaster lies. The VIth Corps was detached from the rest of the Army, and should have received the most careful Instructions, containing all the available information about the enemy and the task it was to carry out. It should then have been given freedom to carry out its task in the manner best suited to the ground and the circumstances of the moment. Samsonoff violated all the principles appertaining to the issue of Instructions. He gave no information about the enemy. He indicated Rastenburg as the front from which danger was to be expected, whereas he would have been better advised to employ such wording as the following: "Some three Corps of the enemy withdrew Westward from the line Benkheim–Gumbinnen during the 21st, after an indecisive battle with the First Army. Some of these troops may withdraw on Königsberg, but the remainder will certainly move Westwards towards Allenstein." The task of protecting the right of the Second Army could then have been carried out by the VIth Corps after proper reconnaissance had been made. Samsonoff definitely tied the 4th Cavalry Division to Sensburg, and himself ordered the direction of their reconnaissance. This order, in fact, discloses the general idea about Cavalry reconnaissance in the Russian Army. The main body of the Cavalry was usually maintained far back, and to a flank, whilst reconnoitring bodies were sent forward to long distances. There was no conception of sending forward the main body of the Cavalry to back up their own reconnoitring parties. As a consequence, continuous touch with the enemy was never maintained, and if the reconnoitring parties did get into touch with the enemy they never managed to get their news back in time. There was much loss of effort in the relief of reconnoitring parties, and touch was lost for long periods. Had the VIth Corps been allowed to arrange the disposition of its own Cavalry, one can imagine the 4th Cavalry Division being employed near Rössel with orders to find out the direction of the enemy's march across the front of the Corps. Any such thing as a complete surprise as actually occurred on the 26th would then have been impossible.

The Commander of the VIth Corps and the Commander of the 4th Division were afterwards dismissed from their appointments for incompetence.

In a vain attempt to guard the right flank of the Army, Samsonoff, at midnight on the 27th/28th, issued an order to the VIth Corps to occupy Passenheim, but on the morning of the 28th a Dragoon reached him from the Chief of the Staff of the Corps, telling him that the Corps was now South of Ortelsburg.

This elicited a message to the VIth Corps by the returning Dragoon :—

"Hold on to the region of Ortelsburg at all costs. The success of the advance of the XIIIth and XVth Corps depends upon the steadfastness of your Corps. Send the 4th Cavalry Division against the troops opposing you. The Division should pursue with energy. Support them as much as you can."

The VIth Corps Commander at 10 P.M. on the 28th ordered the 16th Division to Lake Waldpusch by 3 P.M. on the 29th, with a view to occupying Ortelsburg on the 30th. The shattered 4th Division was to continue its withdrawal to Fürstenwalde on the Russian Frontier.

This was hardly an adequate reply to the order, "Hold on to the region of Ortelsburg at all costs."

The 4th Cavalry Division was ordered on Passenheim.

On the night of the 29th/30th, when the remnants of the Second Army were being destroyed in the forest of Grünfliess, only 16 miles away, the VIth Corps was passively resting with a complete Division near Ortelsburg, the Corps Commander reporting :—

"I am going to attack Ortelsburg with the 16th Division, and am moving the 4th Division to Radzienen, as I have news of the enemy at Willenberg."

This collapse of a Corps and a Cavalry Division upsets calculations as to the delaying power of a modern force of all arms. The Germans had suffered severely at the Battle of Gumbinnen, they had withdrawn by forced marches—the advanced troops of the XVIIth Corps covering 160 miles in eight days,—and their march Southwards had been hampered by hosts of refugees crossing the line as they fled

Westwards. The Russians had covered 132 miles in 11 days, an average of 12 miles a day, and they had had no casualties from enemy action. The crux of the matter, altogether outside the bad tactical leading on the 26th, was that the Russians were unfed. The Germans were properly fed. An underfed man will never be a match for a fed one.

With the VIth Russian Corps driven from the field, the XVIIth Corps continued its march Southwards to the line Gr. Schiemanen–Omulefofen, thus closing the Northern exits of the Forests of Grünfliess and Kaltenborn. The I[st] Reserve Corps swung away on the 27th towards Allenstein to cut in on the Russian centre.

4. The Attack of the Russian Left (Sketch F).

By the night of the 25th August the I[st] Corps was still short of some of its Artillery and Ammunition Columns, and its Commander, Von François, twice protested against being ordered to attack, but Ludendorff had no time to waste and would hear of no delay. So determined was Von François' opposition to attacking that Hindenburg paid him a personal visit at Montowo on the 25th, and impressed upon him the necessity for haste.

The order from the Eighth Army which set the German attack in motion was issued from Riesenburg at 8.30 P.M. on the 25th August, and ran as follows :—

"Definite news has arrived that portions of the

I[st] Russian Corps are dug in about Bergling–Grieben in front of the XXth Corps.

"The I[st] Corps will seize the heights of Seeben at 4 A.M. to-morrow with its left flank, preparatory to attack at the latest by 10 A.M. from Seeben in the general direction of Usdau.

"Muhlmann's Detachment will remain under the I[st] Corps.

"The reinforced XXth Corps will maintain its position, and will support the attack of the I[st] Corps by an attack on its right in the direction of Gr. Grieben–Jankowitz. It will be prepared to advance along the whole front with its main force on its right.

"The 3rd Reserve Division will move forward against Hohenstein.

"I will be at Riesenburg till 7 A.M., and will then proceed to the Eastern exit of Löbau.

HINDENBURG,
Commander-in-Chief."

The I[st] Russian Corps was occupying a position along the high ground Usdau–Meischlitz–Ruttkowitz, with advanced troops pushed forward to Seeben and Grallau. The 15th Cavalry Division was watching Lautenburg, and the 6th Cavalry Division was at Gr. Lensk. On the right of the I[st] Corps there was a gap of about seven miles, the 2nd Division of the XXIIIrd Corps being near Skottau.

Of the reinforcements allotted to the Russian left flank, the 3rd Guard Division, 9th Sapper Battalion, the 1st Rifle Brigade, and the Heavy Artillery, none were within supporting distance.

On the actual morning of the 26th, the opening

day of the Battle of Tannenberg, Samsonoff wired to the I[st] Corps :—

"Your task is to protect the rear of the Army on the left flank, and this task must be carried out at all costs. The Army Commander is convinced that even a greatly superior enemy is not in a position to break the resistance of the famous I[st] Corps, upon whose action the success of the operations against Allenstein and beyond depends."

The Corps Commander could have had no doubt as to the importance of his task.

News had been accumulating that the Germans were in strength at Lautenburg, but the two Russian Cavalry Divisions had merely reported and withdrawn without fighting. There was no news as to the arrival of the I[st] German Corps in the vicinity of Hartowitz, and the Russians were in complete ignorance of any attack coming from this direction.

The I[st] Russian Corps had been in position since the 24th. The Corps Commander had 50 Squadrons of Cavalry at his disposal, against 9 on the German side, and yet a whole German Corps had been allowed to detrain within a day's march of his main position, and deploy in front of it without being detected. This is a typical example of the way in which the Russian Cavalry was employed. There was no attempt at reconnaissance to the front of the position, and the Cavalry was tied to the protection of the left flank. When opposed by Infantry, the Cavalry invariably retired, so that the protection they afforded to the flank was purely illusory.

The Russian Corps Commander decided that he could best fulfil his task of protecting the left flank and rear of the Army by attacking the only enemy force he knew of in his front, that at Lautenburg. To find the troops for this attack, he took a Brigade of Infantry from his right and moved it down to the left. Not only did he weaken his right by still further increasing the gap between himself and the XXIIIrd Corps, but he gave the Brigade no time in which to reconnoitre the ground over which they were to advance, and they only reached their position of deployment in the early morning of the 26th.

The Ist Russian Corps, supported by two Cavalry Divisions, was about to be attacked by an equal force of Infantry in front, and in the right flank by a strong Brigade, and both these attacks were to come as a complete surprise. The Russians were about to attack the German right.

On the morning of the 26th Von François attacked half-heartedly, with deliberate intention. His Artillery was not up, and he had no intention of expending his Infantry. Although the Russians were easily driven from their forward positions at Seeben and Grallau, he issued no orders for the attack on Usdau. The XXth Corps flank attack went no farther than Gr. Grieben, when it was realised that the Ist Corps was not going to attack according to orders.

The Russian troops were in no condition to fight, and the withdrawal from Seeben became a rout, so that the Litovski Guard Regiment, which was passing in rear to join its own Corps, the XXIIIrd, had to be thrown in to hold the heights of Usdau.

On the extreme right of the German line, Von

Muhlmann's Detachment, as it moved forward, ran into the Russian Brigade advancing to attack. They occupied and held Heinrichsdorf, bringing the Russian advance to a standstill.

This Russian Advance against Lautenburg had available 8 Battalions, 4 Batteries, and 22 Squadrons of Cavalry to oppose Von Muhlmann with 6 Battalions, 4 Batteries, and 1 Squadron, and there were a further 22 Squadrons well placed to cut in upon his immediate rear. Both the Russian leading and the state of the troops were at fault. The Corps Commander issued orders to the two Cavalry Divisions to co-operate with the attack, but did not place them under the orders of the Divisional Infantry Commander. The latter allowed the Infantry Brigade Commander to direct the fight, and no orders were passed to the Cavalry to inform them of what the Infantry were doing. The Russians made no progress, and the 6th Cavalry Division having dismounted its men, became much mixed up with the Infantry.

The 15th Cavalry Division remained watching Lautenburg, which was unoccupied, and finally withdrew across the Frontier to Zielun in the evening.

The failure of the Russian Advance against the German right enabled Von François to prepare quietly for his attack next day. Had the Russians retained any fighting value, they might have overwhelmed Von Muhlmann and punished Von François for his half-hearted attack.

During the day the Ist Russian Corps Commander had visited his right near Usdau, and had been prevailed upon to allow the advanced troops to be withdrawn along the Front. He made no attempt to impress upon his subordinates that the holding of the position was vital to the safety of the Army.

By the night of the 26th/27th all Von François' Artillery had come up, and he was ready for his great blow.

His orders for the 27th were as follows :—

"No decision was obtained to-day. The enemy has apparently strengthened the heights of Usdau. The attack on this place will be carried out to-morrow (27th).

"1st Division will contain the enemy at Usdau with a portion of its troops and the Heavy Artillery, while the remainder of the Division will attack from the N.W.

"The 1st Grenadiers will be placed at the disposal of the Division. They should be in the neighbourhood of Gr. Grieben. (*Note.*—This regiment had been lent to the XXth Corps, and now returned to the Ist Corps.)

"The 2nd Division will attack Usdau from the direction of Fichtenwalde, securing itself to the South.

"Muhlmann's Detachment will maintain its position at Heinrichsdorf, preventing any enemy move from Soldau.

"The Artillery will commence firing on the Usdau front at 4 A.M. At 5 A.M. the converging attack of the 1st and 2nd Division will be pushed home.

"I will be on the Tautschken–Seeben road by 5 A.M.

V. FRANÇOIS."

No mention is made of the Detachment under Von Schmettau from the XXth Corps, which was to attack the right of the Ist Russian Corps.

The Russian position from Usdau through Meischlitz to Ruttkowitz is of great strength. The field of fire is excellent, and the slopes leading up to the main position absolutely without cover. There was ample room within the position for the concealment of Artillery close up to the line, and there was excellent observation right up to Seeben and Grallau. Ample time had been available to strengthen the line since the 24th.

For the 27th, a Regiment of Cavalry and a Horse Artillery Battery from the 6th Cavalry Division had been moved up to Rauschken to fill the gap between the Ist and XXIIIrd Russian Corps.

On the 27th, the German Artillery opened with a hurricane bombardment against the Mühlenburg, the key of the Russian position just North of Usdau. The Russians began to leave their trenches without waiting for an Infantry attack, and the whole line gave way under the effect of the Artillery fire. The movement soon became a rout, and by 10 A.M. the field was clear, the Ist Corps having ceased to exist as an organised body. The 6th Cavalry Division was also involved in the rout, and no attempt was made to cover the movement. The 15th Cavalry Division remained inactive at Zielun.

The Battle of Usdau, if it can be so called, had been won by the German Artillery. The unfed Russians had no stomach to withstand its effect. Von François describes the Russian trenches as being full of dead when he saw them on the 28th. The breaking of the I^st Corps was the first step in the final disaster about to overcome the Second Army. Its complete collapse came as a surprise to the Army Commander, Samsonoff, who did not realise to what state the men had been brought through lack of food.

By 10 P.M. on the 27th all that remained of the Corps was a rear-guard of five Regiments from various Divisions, with five Batteries of Artillery under the command of General Sirelius, the Commander of the 3rd Guard Division (XXIIIrd Corps). This body took up a position along the railway just North of Soldau.

During the night of the 27th/28th, Von François had located the position of his various detachments, and ascertained that the enemy was making a stand North of Soldau.

At 6.40 A.M. on the 28th he issued the following orders :—

"1. The 1st Division and Schmettau's Detachment from the line Fylitz–Klenzkau–Schönwiese to advance on Soldau.

"2. 2nd Division to remain in Reserve at Gr. Tauersee.

"3. Muhlmann's Detachment to seize Pierlawken and maintain touch with the 1st Division.

"4. The Cavalry of the 1st Division and of Schmettau's Detachment, strengthened by a Battery and some cyclists, to move *viâ* Gr. Koslau on Neidenburg."

His intention was to drive the remains of the I^st

Corps South and away from the rest of the Russian Army, so that the intended flank attack would not be interfered with. It soon became evident that the Russians had evacuated Soldau during the night of 27th/28th, and were retiring in disorder.

There was nothing to oppose the victorious Germans. The Russian left flank lay completely exposed.

All had gone well with Von François, but the XXth German Corps was in difficulties to the North. An urgent wire from Ludendorff ordered Von François to move his Reserve Division, the 2nd, from Tauersee *viâ* Kramersdorf–Wiersbau to Rontzken, and so cut in on the flank and rear of the Russians opposing the XXth Corps.

A further Army wire gave the general direction of the advance of the I[st] Corps as Lahna. Evidently Ludendorff was nervous about the XXth Corps, and wished to close Von François to the North.

But Von François, ever independent, had his eye fixed upon Neidenburg and the main road to Willenberg, and he deliberately altered the line of his advance.

At 4.30 P.M. he issued the following orders (Sketch H) :—

"To-day, 28th, the following marches will be carried out :—

Von Schmettau's Detachment—Muschaken.
1st Division—Neidenburg.
2nd Division—Grünfliess.
All available Cavalry South of Neidenburg to Willenberg.

"On the 29th :—

Von Schmettau's Detachment—Willenberg.
1st Division—Muschaken.
2nd Division—Grünfliess."

These marches were completed according to orders. The marching of Von Schmettau's Detachment of Garrison troops was particularly good. They were at Bergling on the morning of the 27th, and had covered 40 miles by the afternoon of the 29th, the march being across country with many halts to deal with the straggling Russians.

Von François had cut right through the flank of the Russian Army, and was proceeding to block the line of the railway from Neidenburg to Willenberg.

Disobedience of orders by a subordinate is always a difficult matter to excuse. Never an easy subordinate to control, Von François deliberately disobeyed the Army order to direct his march on Lahna. He ignored the situation of the XXth Corps on his left, and gambled against the danger of a Russian attack against its right flank which he had exposed. He even made no effort to inform Ludendorff of his action in contravention of a definite order.

Had he obeyed his orders there would have been great disorder at Neidenburg on the 30th August, when the Russians attacked under Sirelius, and it is more than probable that large portions of the Second Russian Army would have escaped envelopment on the 29th and 30th. It was Von François' contempt for the fighting condition of the Russians that made him push forward so boldly to Neidenburg and Willenberg. His long extended line of picquets successfully resisted the half-hearted attacks of the exhausted Russians trying to force their way out of the forests of Grünfliess and Kaltenborn, but against troops in better condition his position would have been perilous.

To the disobedience of Von François, Ludendorff undoubtedly owed the magnitude of his victory on the succeeding days.

5. The Attack on the Russian Centre (Sketch G).

In accordance with Order No. 6353, the Centre of the Second Russian Army continued its advance on Allenstein–Osterode during the 26th.

The XIIIth Corps advanced in two columns, the 1st Division *viâ* Kurken to Stabigotten and the 36th Division *viâ* Wuttrienen to Alt Allenstein, where they halted for the night. No contact was obtained with the enemy.

Opposite the XVth Corps the enemy was reported to be holding a line from Nadrau–Paulsgut–Mühlen. The Corps made good progress, driving the enemy in front of it, and by evening reaching the line Grieslienen – Hohenstein – Königsgut – Lichteinen–Paulsgut–Ganshorn to just East of Mühlen, which latter place they failed to take after stubborn fighting owing to the non-arrival of the 2nd Division on their left. The XVth Corps fought well on this day, despite the weariness of the men.

The 2nd Division (XXIIIrd Corps), temporarily under the orders of the XVth Corps, had been ordered to attack Mühlen in concert with that Corps.

They started well by driving the enemy out of Thurau, but were immediately counter-attacked from the direction of Faulen and Logdau and subjected to heavy fire from Artillery of large calibre, evidently guns from the Vistula fortresses. The two Brigades of this Division fell back in some disorder on either side of the Kownatken Lake to the line Januschkau–Lippau. The Revel Regiment on the extreme left was taken in flank through lack of reconnaissance and annihilated, losing actually 51 officers and 2,800 men in killed, wounded, and prisoners. The Germans followed up closely, and only darkness saved the 2nd Division from complete disaster. The attacking Germans were about a Brigade of the 41st Division (XXth Corps).

The night of the 26th/27th must have been an anxious period for the German Army Headquarters. The I^st Corps had done little on the 26th, and the I^st Reserve Corps, after its success against the VIth Russian Corps near Bischofsburg, although ordered to close towards Allenstein, was still some distance away. The XXth German Corps had been sorely tried, and its units were much mixed up with those of the 3rd Reserve Division and Von Unger's Garrison Detachment. They had had to give way in front of the advance of the XVth Russian Corps, and the men were very tired. The right of the XXth German Corps had indeed gained a distinct success, but it had been drawn away from the remainder of the Corps into a difficult Lake country. The only reinforcement available was Von der Goltz's Landwehr Division, which could not be expected to begin detraining at Biessellen, West of Allenstein, until the morning of the 27th, and they had already been seven days in the train.

Two Russian Corps at least were advancing against the line Allenstein–Mühlen, and there was little with which to oppose them.

Time was essential to allow the German I^{st} Reserve and I^{st} Corps to close in on either flank of these two Russian Corps, and it is typical of the German Command that they ordered the XXth Corps, with its attached troops, to attack on the 27th to gain that time.

The Russians continued their advance on the 27th according to orders.

At 8.30 A.M. the Commander of the XIIIth Corps wired :—

" The Germans are carrying out their War Game. I shall keep that in view to-morrow in my further movements. Order the Intendant to continue the étapes to Jedwabno and further. I am in great want."

At 9.30 A.M. :—

" Allenstein is by reports weakly occupied. I will leave it to the other Corps. I am going down to help the XVth Corps."

The VIth Corps was expected at Allenstein according to orders, which, as we know, had been cancelled. The Commander of the XIIIth Corps, being unable to get touch with the VIth Corps, sent only one Brigade of his Corps to help the XVth, and moved with the remainder on Allenstein, which he occupied about 4 P.M. He notified the Army that there was no sign of the Germans in the town, and that the VIth Corps had failed to arrive. He proposed to move on Osterode on the 28th.

The men of the XIIIth Corps evidently found liquor in the town, and there was some looting and disorder, which is not surprising considering the famished state of the Corps. No attempt was made to organise proper security or to cut the railway lines, with the result that a German Armoured Train arrived and shelled the camp of the 1st Division, which was S.W. of the town.

The Brigade of the XIIIth Corps, sent to help the XVth Corps, lost its way in the wooded country, and finally spent the night at Schwenteinen, without having gained contact with the enemy or having found the right flank of the XVth Corps.

At the moment that the XIIIth Corps was resting at Allenstein on the night of the 27th/28th, reporting no enemy in the vicinity, the Ist German Reserve Corps was closing in upon it from the N.E. This Corps lay on this night some 10 miles East of Allenstein between Wartenburg and Passenheim.

The XVth Corps, advancing Westwards between the Hohenstein–Reichenau road and the village of Mühlen, ran into the units of the XXth German Corps advancing Eastwards. Fighting was particularly fierce and casualties heavy on both sides. The Russians succeeded in occupying a line running from just East of Heidemühl and Dröbnitz through a portion of the Wahlsdorf forest to just West of Mühlen, which they captured. They had crossed the difficult ravine of the R. Drewenz on their left, and had inflicted heavy casualties on the Germans, but their advance had been held up, and the men were nearly exhausted. The German advance had served its purpose in gaining time.

On the left of the XVth Corps, the 2nd Division, which had been driven back in two detachments to Januschkau and Lippau, was again driven back on the 27th. The right detachment was rallied in Frankenau by the Commander of the XXIIIrd Corps, who threw in the Kexholmski Regiment of the 3rd Guard Division, which had just come up from Neidenburg. The left detachment crowded in disorder into Neidenburg about mid-day, and the Staff of the Second Army turned out to reform them. A Staff Officer reports upon them as follows :—

"The men were terribly exhausted, having very few rounds in their possession. They had been three days without bread or sugar."

The Commander of the Estlyandski Regiment reports :—

"I had 450 men. For two days they had received no rations, and none of the supplies had come up. The men were so worn out that they saw an enemy in every body of troops."

The defeat of the 2nd Russian Division had exposed the whole of the left flank of the XVth Corps.

The night of the 27th/28th must have been a relief for German Army Headquarters. The position had completely altered after the fighting of the 27th. They now knew that the Russian left had been driven in disorder from their positions by the Ist Corps, and that the Ist Reserve Corps was closing in on Allenstein.

Late on the evening of the 27th, Ludendorff issued the following order :—

"The reinforced XXth Corps, Von der Goltz's Landwehr Division, Ist Reserve Corps, and XVIIth Corps, will envelop the Russian XIIIth and XVth Corps."

It will be remembered that Ludendorff had also ordered the Ist Corps to close in to the North by directing his General advance on Ladna, but Von François had disobeyed his orders.

Samsonoff had moved up the Headquarters of the Second Army to Neidenburg on the 27th, and almost the first sight that met his eyes was the disordered Brigade of the 2nd Division streaming into the town. At midnight on the night of the 27th/28th he issued the following order :—

"To-day, the 27th August, the Ist Corps has been fighting on the line Usdau–Heinrichsdorf–Gr. Lensk, beating off the German attacks.

"The 2nd Division, after a desperate battle at Januschkau, has taken up a position West of Frankenau.

"The XVth Corps has occupied the village of Mühlen and the Forest of Wahlsdorf.

"The XIIIth Corps is in touch with the XVth Corps, and has occupied Allenstein.

"On the 28th, the Ist Corps will maintain its position in front of Soldau at all costs, to protect the left flank of the Army. The 2nd Division and portions of the 3rd Guard Division will maintain their positions West of Frankenau at all costs.

"The XIIIth and XVth Corps, under the command of General Martos (XVth Corps), will move at once in the direction of Gilgenburg–Lautenburg, with the object of attacking in flank and rear the enemy advancing against the XVth Corps and 2nd Division.

"The VIth Corps will move to the region of Passenheim."

This order hardly represents the true situation, even as Samsonoff knew it that night. The Headquarters of the Ist and XXIIIrd Corps, during the 27th, were at Soldau and Salusken, 4 and 14 miles respectively from Neidenburg. Liaison by motor or horse could have been carried out in very few minutes, when the true situation would have been realised. It is possible, however, that Samsonoff realised that the troops on his left had been fighting badly, and did not wish to give a bad impression of the situation to the other troops.

General Knox in his book says that the Headquarters of the Army were of opinion that they were opposed by a general strength of three German Corps, two Regular and one Reserve, and that the position was regarded as hopeful.

It is difficult to visualise the picture that Samsonoff made of the situation at the moment he wrote this order. His right flank Corps had been driven back a considerable distance, and was in some confusion. The 1½ Corps on his left, supported by two Cavalry Divisions, had also been driven back. The right Corps of his Centre had had no fighting, whilst the left Corps of the Centre had been fighting heavily. It would appear that he considered the main enemy strength to lie on his left, and to consist of the XXth German Corps and any reinforcements the Germans had been able to find from the Vistula. The Germans on his right, who had attacked his VIth Corps, were probably a residue from Lötzen, and were escaping so as not to be shut in by Rennenkampf. Samsonoff had no information as regards the troops who had opposed Rennenkampf at Gumbinnen, either from Rennenkampf or Jilinsky. Neither the VIth nor Ist Corps had been able to obtain any identifications which might have given him some indication of the situation. The XVth Corps identifications had all been of troops not engaged at Gumbinnen.

Samsonoff evidently took it for granted that Rennenkampf was in close touch with the Germans he had supposedly defeated at Gumbinnen, and had no fear from the North. He intended to envelop the Germans on his left flank.

Two officer patrols in the XIIIth Corps had indeed reported on the 27th that large forces were approaching Allenstein from the N.E., but this information was not passed on by the Divisional Commander concerned, who took them to be the VIth Corps arriving according to orders. No attempt was made to verify this assumption.

The XIIIth and XVth Corps were to face directly to the South and continue their advance. No attempt was made to co-ordinate the march of these two Corps, this being left to General Martos, at the time heavily engaged along his whole front. General Martos was not even told when to expect the XIIIth Corps or where to find them.

The day of the 28th was to undeceive Samsonoff bitterly as regards his favourable position.

At 5 A.M. the XIIIth Corps left Allenstein for Hohenstein to join the XVth Corps, leaving two Battalions in the town as rear-guard. The Corps commenced deploying at Grieslienen against troops advancing from the N.W. Some time elapsed before it could be ascertained whether these were Germans or not. The Corps was eventually ordered to attack Hohenstein, but found the town in German hands, and by the evening had failed to retake it. A position was then taken up for the night between Grieslienen and Mörken facing N.W.

At about 1 P.M. the two Battalions left behind in the town of Allenstein found Germans approaching the outskirts of the town from the N.E. and E. Two more Battalions from the main body of the XIIIth Corps were sent to extricate them, but during the afternoon these four Battalions found their road to Hohenstein cut, and they were finally driven up against Lake Wulping and forced to surrender. This episode is doubtless the origin of the fable that the Second Russian Army was destroyed in the marshes of the Masurian Lakes.

The German attack on the XVth Corps on the 28th was carefully staged. Von der Goltz's Landwehr Division was to come in from the North *viâ* Sprechan (it was against this Division that the Russian XIIIth Corps deployed at Grieslienen). The 37th Division (XXth Corps) was to attack North of Schwenteinen and Hohenstein, the 3rd Reserve Division between Lichteinen and Königsgut, Von Unger's Detachment against Paulsgut, and the 41st Division (XXth Corps) against Ganshorn from the South.

The left of the XVth Russian Corps made a good fight, driving back the German 41st Division near Ganshorn, and taking prisoner 34 officers and 1,500 men. On the right, however, the endurance of the Russians had reached its limit, and the non-arrival of the XIIIth Corps destroyed their remaining power of resistance. Units began to break all along the line, withdrawing in disorder towards Nadrau, South of Lake Maransen, closely followed by the Germans.

The XVth Russian Corps had borne the brunt of the fighting which had taken place in the Second Army. They were the only troops which had shown themselves capable of opposing the Germans, and their success appears to have been due to the energetic personality of their Commander, General Martos.

The 2nd Russian Division, which had been holding a position just West of Frankenau, supported by units of the 3rd Guard Division, was attacked on the 28th by the Brigade of the XXth German Corps, which had driven them back during the two previous days. The stiffening supplied by the 3rd Guard Division was sufficient to stop the Germans. The Kexholmski Guard Regiment took some 1,000 prisoners on this day.

The Germans came on carelessly after their previous successes, and were surprised by the unexpected resistance. Such episodes in a pursuit will often occur, unless proper attack formations are maintained.

The Commander of the XXIIIrd Corps, General Kondratovitch, who was in command of the Frankenau operations, appears to have lost his head completely. He attempted to move down by car to Neidenburg to organise the defence of that place, but found the road cut by the Germans. He then attempted to return to Frankenau, but found himself cut off from there also. He turned South, and succeeded in dashing through the German lines to the West of Neidenburg by throwing German cloaks over himself and the driver. He was dismissed later by the Generalissimo for having deserted his troops.

It was on the morning of the 28th that Samsonoff became aware of the complete defeat of his Ist Corps on the left flank, and he left Neidenburg at an early hour, sending this, his last wire, to the N.W. Front :—

"The Ist Corps was heavily defeated yesterday evening, and was withdrawn to Illovo by order of the Corps Commander, General Artamonoff, leaving a rear-guard at Soldau. I am going immediately to the Headquarters of the XVth Corps at Nadrau to take over personal direction of the advancing Corps. I am closing my office, and will be temporarily out of communication with you."

As he left Neidenburg he encountered streams of stragglers moving Southwards. He turned up the road to Jedwabno instead of that to Nadrau owing to news that Germans were already across the latter. At 9.30 A.M. he received the full report of disaster to the VIth Corps on his right. He now began to realise the critical situation of the Army. The Germans were in strength on his right flank as well as on his left. He sent a last order to the VIth Corps to throw in their Cavalry Division, and at least to hold on to Ortelsburg. At 11 A.M., Samsonoff, with seven of his Staff Officers, mounted horses taken from the Cossack escort, and rode away to find the XVth Corps Headquarters. General Knox gives a vivid description of this scene in his book. Samsonoff would not allow a foreign officer to accompany the Staff owing to the critical situation. The Army Commander was quite calm, and as he was riding away said, "The enemy has luck one day, we will have luck another."

6. Final Destruction of Russian Centre (Sketches G and H).

(i) *XVth and XXIIIrd Corps.*

A column consisting of the remnants of the XVth and XXIIIrd Corps, with a Brigade of the 6th Cavalry Division, commenced its withdrawal by darkness on the 28th. Rear-guard actions were fought at Michalken, Salusken, and Radomin towards Grünfliess. A Regiment of the Guard, with units of the 2nd Division on its right, and of the XVth Corps on the left, held on to Radomin until 3 P.M. on the 29th, when the 2nd Division began to give way. About 7 P.M. on the 29th the advanced portions of the column reached a point about a mile from Muschaken, and a determined attempt was made to break through.

The Germans from near the road and railway held all the exits from the forest under fire from guns and machine-guns, and successfully prevented the exhausted Russians from coming to close quarters.

The clearings South of the forest are absolutely without cover, and the exits are clearly in view from the line of the road. The Russians were impeded by large bodies of exhausted transport, which attempted to debouch from the narrow exits, mere sandy tracks, but the only means of getting out of the forest. In many cases, waggons and horses were piled up one on top of another by the German Artillery fire, adding to the confusion in the growing darkness. Finding the way to the South blocked, the remnants turned Eastwards once more, towards Rettkowen and Kannwiesen, where a consultation of Senior Officers took place. It was decided to continue on Willenberg, but the men were exhausted and could go no farther. The majority surrendered later between Rettkowen and Saddek without further fighting.

Had there been a determined effort to break through the line of German picquets as soon as darkness came down, doubtless a large number of men would have safely escaped. But it must be remembered that it is easy to lose the sense of direction in a forest at night, and even if there had been compasses, few of the rank and file could have read the maps printed in Latin characters with which the Russians were provided, and there are few direction posts to be seen anywhere on the forest tracks.

Two organised bodies, belonging to the XVth Corps, under resolute Commanders, did succeed in forcing their way through, the course they followed being as follows: One and a half Regiments of the 6th Cavalry Division, with the 11th Horse Artillery Battery, under General Baron Stempel, reached

Kannwiesen on the night of the 29th. They were here joined by General Torklus, the Commander of the 6th Division, and his Chief of Staff. All the roads round Kannwiesen were blocked with carts three deep moving slowly in various directions in the heavy sand. During the night a Sapper Company joined up, and the column approached Willenberg, only to find the enemy in occupation. Throwing out a screen towards the town, the column passed Southwards without being fired at *viâ* Kl. Pivnitz and Opalenitz, finally reaching Chorzele on the morning of the 30th. Some 15 officers and 1,230 men of the 31st Alexis Regiment, under Lieut.-Colonel Suchachefski, reached Saddek, where they found the roads blocked by an inextricable mass of transport. The Commander formed an Advanced Guard of his best men, which he personally led Southwards towards Puchallowen. They succeeded in surprising two German picquets, and the whole column passed safely across *viâ* Sachen to Przasnysz, which was reached on the night of the 30th/31st.

(ii) *XIIIth Corps.*

The XIIIth Corps received orders to withdraw to Kurken at midnight on the night of 28th/29th. Rear-guard actions were fought at Mörken and Schwedrich. Two Regiments took each other for the enemy, and several casualties ensued. Kurken was reached at 12 noon on the 29th. About 2 P.M. on the 29th the Chief of the Staff of the Corps, General Pestich, returned from Orlau, having seen

General Samsonoff, and delivered the following order :—

"The XIIIth Corps will withdraw to Chorzele, the XVth Corps and portions of the XXIIIrd Corps to Janowo. The movement on Muschaken will be covered by units of the XVth and XXIIIrd Corps on the line Lahna–Bartoschken. In order to facilitate the withdrawal of these units, the movement of the XIIIth Corps to Muschaken will be pushed as fast as possible with the object of taking up a position facing Neidenburg by dawn on the 30th. The general direction of the withdrawal of the three Corps the Commander-in-Chief places in the hands of the Commander of the XIIIth Corps. The Headquarters of the Army is moving to Janowo."

Samsonoff had found he could not direct the Corps himself, and he was now trying to regain a point from which he could again get into communication with the N.W. Front. Having failed to find General Martos (XVth Corps), he now handed over the hopeless task to General Klueff (XIIIth Corps).

The XIIIth Corps Column reached Jablonken by 7 P.M. After a short halt the march was continued on Wickno and Kommusin. By darkness, the column ran into the massed transport of the XVth and XXIIIrd Corps withdrawing Eastwards.

Klueff realised that the situation had now altered, and he directed his column on Kaltenborn and Wallendorf. The men were utterly exhausted. They had covered 42 miles in the last 40 hours without rations of any sort or kind. Horses were unfed and unwatered, and the units were inextricably mixed.

The sufferings of both men and animals, dragging slowly along in the dark in the heavy sand, from which there came a fine dust to further increase their agony, must have been unendurable. At Kaltenborn the Advanced Guard ran into machine-gun fire coming from the North, from units of the XVIIth German Corps. The main body turned South towards Muschaken, but were again met by machine-gun fire, probably from the retreating XVth Russian Corps. They again turned North to Wallendorf, which they reached at 11 A.M. on the 30th. Here they surrendered to the German clearing parties, which had moved up into the forest from the South. The Advanced Guard actually captured 4 guns from the Germans in the forest somewhere South of Omulefofen, but withdrew to Wallendorf, where they ran into Germans coming from the South. They finally reached Uszannek, where they entrenched themselves in a circle. All through the 30th and the night of the 30th/31st this body of Russians fought steadily on until there was no more ammunition left. This last fight must have been a magnificent one on their part, famished and exhausted as they were.

The little cemetery at Uszannek contains the graves of many Germans and Russians, testifying to their undefeatable courage.

7. General Samsonoff's Movements.

The movements of General Samsonoff, as far as can be ascertained, were as follows :—

He reached Nadrau on the 28th, being met with

reports that General Martos (XVth Corps) had been killed. He then went to Orlau, where he met the Chief of Staff of XIIIth Corps, and issued him the order which has been quoted. He had made up his mind to ride to Janowo, and finding the direct road blocked, he and his Staff eventually reached Saddek. Here they struck the mass of transport, making it impossible to use the roads any more, so it was decided to let the Cossack escort shift for themselves, while the journey was continued on foot. They had a compass, but no map, and finally the matches by which to read the compass gave out. The party stumbled on, often hand-in-hand, through the forest until the forester's hut near Pivnitz was reached. Samsonoff, who suffered from a weak heart and was in no good physical condition, had covered six miles and could do no more. Muttering "the Emperor trusted me," he slipped aside in the darkness, and shortly afterwards his Staff heard a shot fired from a revolver. An hour's fruitless search brought no result, and the Staff reluctantly decided to continue on its way, certain that the Army Commander had committed suicide in his despair. On the morning of the 30th they struck two Squadrons of Cavalry, and succeeded in passing the Frontier with them.

8. The Efforts of the Headquarters N.W. Front to Extricate the Second Army.

Jilinsky's order to Rennenkampf on the 26th August, giving him fresh objectives, had no indication in it of a necessity to hurry. It was late on the 27th before he wired to Rennenkampf to push forward to the help of Samsonoff. The Battle of Tannenberg had been in progress for two days already.

Samsonoff had cut himself off from Jilinsky on the 28th. From that moment onwards only sheer luck could direct any fresh troops to his assistance. Jilinsky appears to have realised that the Germans had broken Samsonoff's flanks, and that there was grave danger of the centre of the Second Army being enveloped. Some sort of energy in the crisis appears to have been infused into the Headquarters of the N.W. Front, for the orders and telegrams are more concise and definite than usual.

(i) *Russian Left.* (Sketch H.)

General Artamonoff, Commanding the I[st] Corps, was dismissed, and General Sirelius of the 3rd Guard Division put in his place. Sirelius was ordered to collect all available men and attack Neidenburg at once. The order reached its destination, and during the afternoon of the 29th two Regiments of the 3rd Guard Division with seven Battalions of the 1st Rifle Brigade, and certain Artillery units, were collected. The whole force amounted to about a Division. With these comparatively fresh units the attack on Neidenburg was commenced with energy. Other units were collected on the 30th from the débris of the I[st] Corps, but were too exhausted to move forward.

The I[st] Russian Corps had withdrawn so precipitately from the field that Ludendorff left them out of the battle picture. He had therefore issued orders for the I[st] German Corps to draw Northwards to Jedwabno for the commencement of the advance against Rennenkampf, who was now approaching from the North.

At a few minutes before 9.30 A.M. on the morning of the 30th a message was dropped in the square at Neidenburg, giving Von François warning of the coming attack. It was short and to the point:—

"Aeroplane A. 29. Lieutenant Hesse. Route Eylau – Soldau – Mlawa – Neidenburg. 30th August, 9.15 A.M.

"To Commander, I[st] Corps.

"Column of all arms moving from Mlawa on Neidenburg. Head at Kandien, tail one kilometre N. of Mlawa. Time, 9.10 A.M. A second column moving from Stupsk on Mlawa. Head at E. entrance Mlawa, tail Wola.

"Time, 8.45 A.M.

KÖRNER, *Observer*."

This was the Russian attack under Sirelius coming in.

The German Commander's action was typical. There was no thought of an abandonment of his position along the Willenberg road at Muschaken and Puchallowen, and so a possible complete victory. There was only the thought of an immediate counter-attack to gain time. There was no time to issue orders in writing, so three officers were sent off in motors to order the attack verbally.

The first to Major Schlimm, Commanding the

Detachment South of Neidenburg. This officer was found to be organising a defence already, along a small line of hills on a level with Berghof. All the Russian prisoners taken South of Neidenburg were streaming Northwards under a small escort. Schlimm was ordered to hold on at all costs.

The second to the 2nd Division near Grünfliess, with orders to advance with all available men *viâ* Gregersdorf and Sagsau to attack the enemy's right flank.

The third to Von Muhlmann at Soldau, with orders to advance *viâ* Saberau to attack the enemy's left flank.

Headquarters of the Corps were moved out of the town to the North, the last motor leaving the town as the first shells began to drop in the square.

Here we see the clear-cut idea which is always present in Von François' tactical dispositions, the holding of the enemy in front and an attack against one or both flanks by forces directed from a distance.

At 1 P.M. Von François received the news that the Commander of the XVth Russian Corps (Martos) had just been captured with 25,000 men. His movement along the Willenberg road had already achieved great success. He faced the coming attack with all the more determination, since now Ludendorff had also received Körner's aeroplane message, and was sending nearly five Divisions Southwards to his support.

The Russian attack came gallantly on, and after fierce fighting, the town was taken about 9 P.M. on the 30th. General Sirelius took up a position in a semi-circle just North of Neidenburg and dug himself in during the night.

On the morning of the 31st he received definite news of the destruction of the Russian centre. He also found himself about to be assailed in front and on both flanks by superior forces. He ordered an immediate withdrawal. The withdrawal on the 31st and the attack on the 30th both seem to have been carried out with skill. Nevertheless, he was adjudged to have acted improperly on this day, and previously for withdrawing the rear-guard from Soldau, and was dismissed from his Command.

(ii) *Russian Right.* (Map II.)

At 11 A.M. on the 30th, Jilinsky ordered the VIth Corps to attack Willenberg. The Corps received the order while commencing half-hearted operations against Ortelsburg. An attempt was made to break off these operations, but it was not till 11 P.M. on the 30th that the 16th Division reached Lipowitz, and the 4th Division Borken. The 4th Cavalry Division reached Grammen in an attempt to penetrate to Passenheim. On the 31st the news of the disaster to the XVth and XIIIth Corps arrived, and the VIth Corps was ordered to cross the Frontier into Poland.

(iii) *Russian Centre.*

In desperation at the lack of news from the centre of the Second Army, Jilinsky wired to Rennenkampf to send forward Cavalry to ascertain Samsonoff's whereabouts. The story of these Cavalry operations is told in a later Chapter.

On the 31st the position of the Second Russian Army is thus described by the Headquarters N.W. Front in a wire to the Generalissimo :—

"There is no information about the situation of the Corps of the Second Army on the evening of the 31st. It can be assumed that the Ist Corps is fighting at Neidenburg, and that the VIth Corps is between Willenberg and Myszyniec. There is no news of the XIIIth Corps. Isolated men of the XVth Corps are arriving in batches at Ostrolenka. A portion of the XXIIIrd Corps is fighting with the Ist Corps, a portion was with the XVth Corps. The remnants are assembling at Przasnysz."

General Postofsky, Chief of Staff of the Second Army, had escaped from the debacle, and had succeeded in reaching Ostrowo, where he received the following telegram from the N.W. Front :—

"As there is no Commander-in-Chief of the Army, you are appointed to command temporarily. You will receive orders as to the collection of troops of the Second Army from the Chief Intendant, General Bobrofsky. According to my information, the Ist Corps is fighting to-day (31st) at Neidenburg. This Corps should be withdrawn to Mlawa, otherwise it may come in contact with superior forces. It is of the utmost importance to get the units organised, fed, and brought into rest. Try and get information as to where the XIIIth Corps is. Isolated men must certainly turn up in large numbers. Keep in connection with the N.W. Front."

On the 31st August, General Scheidemann, Commander of the IInd Corps, was appointed to command the Second Army.

9. Numbers at Tannenberg.

The following table gives a comparison of the opposing forces :—

Russians.

	Battns.	Sqdns.	Guns. Field.	Guns. Heavy.
Ist Corps	24	2	120	..
VIth Corps	29	8	112	..
XIIIth Corps	31	5	116	..
XVth Corps	28	8	120	..
XXIIIrd Corps	20	1	104	..
4th, 6th, and 15th Cav. Divisions	..	72	36	..
6th Heavy Arty. Div.	..	..	..	12
	132	96	608	12

Germans.

	Battns.	Sqdns.	Guns. Field.	Guns. Heavy.
Ist Corps	24	8	144	32
XVIIth Corps	24	8	144	20
Ist Reserve Corps	26	8	144	16
3rd Reserve Division	12	3	36	..
1st Landwehr Div.	12	4	12	16
XXth Corps	25	6	144	23
Vistula Garrisons	32	11	66	21
	155	48	690	128

It will be seen that the Germans were superior in Infantry and greatly superior in Heavy Artillery.

They carried out their great enveloping movement without the help of any Independent force of Cavalry. The apparent precision of their movements and excellent information was due to the rash use made of wireless by the Russians, who either sent messages in clear or in such a simple cipher that they were decoded with ease at German Headquarters.

The actual tally of losses and prisoners on the Russian side has been variably given by the Germans and Russians. The practical result of the fighting from the 26th till the 30th August was the total annihilation of two Corps (XIIIth and XVth), from which there escaped only 50 officers and 2,000 men. The XXIIIrd Corps mustered 3,000 men, most of whom went with General Sirelius in his attack on Neidenburg, and had not taken previous part in the fighting. The VIth Corps was reduced to the strength of about a Division, whilst the Ist Corps was about the same strength, though completely demoralised by their defeat and retreat from Usdau.

The Germans claim that the results of the battle were :—

95,000 Unwounded prisoners.
30,000 Wounded prisoners.
500 Guns.

It is certain that the booty was very large, for nearly a week was required to collect horses and transport along the line from Muschaken to Puchallowen. Over 60 trains left Puchallowen, the horses being driven in herds to improvised camps in the vicinity of Neidenburg.

10. Comments upon the Battle of Tannenberg.

The disaster of Tannenberg was the greatest defeat suffered by any of the combatants during the war. The suddenness and unexpectedness with which it was achieved heightened the strategical effect of the battle upon the war as a whole. Confidence in Russia amongst her Allies was completely shaken. The German nation was raised to such heights of enthusiasm by the exploits of Hindenburg and

Ludendorff that the defeat on the Marne, which came about a fortnight after Tannenberg, passed without much comment amongst the people. For the German Government the victory of Tannenberg was a godsend, diverting attention from their ill-successes elsewhere.

Jilinsky had hurried the Second Army into an advance before it was ready. Samsonoff made no protest when given an impossible time-table of marches, and exhausted his army before he had gained contact with the enemy. He had persistently disobeyed his orders in extending his advance to the Westward, and had thereby upset the timing of the campaign. Jilinsky had weakly acquiesced in the disobedience of his subordinate, and took no steps to adjust the situation by directing the First Army to close in to the Second.

Reconnaissance, both strategical and tactical, had been utterly neglected during the advance, and the Russian forces had been gradually dissipated upon subsidiary objectives until only half of the original army remained as a striking force.

Surprised everywhere, and more than half-starved, only in one Corps, the XVth, did the Russians put up a fight worthy of the name of their distinguished regiments.

Ludendorff might well be proud of his manœuvre. He had destroyed one of the Russian Armies invading East Prussia in the short space of five days (26th to 30th August).

CHAPTER VII.

The German Advance against Rennenkampf: Battle of the Masurian Lakes.

Map II.—The Advance of the Russian Armies.
Map III.—The Advance of German Armies, 2nd to 14th September.

1. Rennenkampf's Forward Movement after Gumbinnen.

The failure of Rennenkampf to come to the assistance of the Second Army has been much discussed. It has been suggested that personal enmity between himself and Samsonoff was the cause of his failure to co-operate, but there appears to be little or no justification for such a suggestion. By all the principles of war, no effort should have been spared to prevent the two Russian Armies from being defeated in detail, and it was Jilinsky's duty to co-ordinate the movements of the Armies. It has already been shown how he had allowed Samsonoff to deviate from his instructions, and it is now time to examine his orders to Rennenkampf.

The First Army had won the Battle of Gumbinnen, in that the Germans had withdrawn after intending to fight a decisive battle. The Russians had been roughly handled, as evidenced by their casualty lists, and there was need to reorganise the flow of their rear services, but sufficient troops were available with which to improve their victory. Five and a half Cavalry Divisions and one and a half Infantry Divisions had not been engaged at Gumbinnen, and were close to the scene of action.

After Gumbinnen, Rennenkampf made no forward move for 60 hours. No touch of any kind was maintained with the retreating enemy. As an example of his regard for time, it is reported that during the night of the 20th/21st, when the news arrived that the Germans appeared to be withdrawing, he turned to one of his Staff, saying that the danger was now over, and that they could go to sleep. From notes made by one of his Staff, it would appear that the Army Commander was of opinion that the Second Army would be late in its timing, and that therefore it would be wrong to push the enemy too rapidly, or they might escape the blow intended to take them in rear. He had been ordered to draw the greatest possible enemy force upon himself, and having done so successfully, he considered that his task was accomplished. Whatever were Rennenkampf's opinions as to the timing of the General Advance, there can be no excuse for his not keeping touch with the enemy in front of him, in order to ascertain which way they went, and in what state of disorder they were. As we know, the Germans were by no means in disorder, and the optimistic reports which reached Rennenkampf emanated from friendly Polish inhabitants. No effort was made to verify these reports, which were passed on as facts to the N.W. Front.

Rennenkampf had failed in one of the first principles of leadership. He had made not the slightest attempt to impose his will upon an enemy already dispirited by an unsuccessful operation. Moreover, he was making no effort to obey his instructions to cut the Germans off from Königsberg. He allowed them to entrain three Divisions within 20 miles of the battlefield of Gumbinnen, and the whole German Army to disappear from in front of him so completely that he had no idea in which direction they had gone.

It was thus not until late on the 23rd August that Rennenkampf set his Army in motion once more. Jilinsky sent him no orders to push on, and made no comment on his failure to keep touch with the enemy.

By the evening of the 23rd the following line was reached :—

Pelleningken–Ischdagen–Darkehmen.

No contact with the enemy was gained, but there were ample signs that the German withdrawal had been a hurried one. The evacuation of the R. Angerapp position, with the visible tokens of haste in the abandoned equipment and half-finished trenches, coupled with many reports of Von Prittwitz's panic decision to withdraw behind the Vistula, all tended to convince Rennenkampf of what he wished to believe—that he had thoroughly defeated the enemy at Gumbinnen.

During the next few days the march continued its leisurely way, the Cavalry well to a flank, and never in front of the advancing line.

24th. Kauschen–Trempen–Skirlyak.
25th. Wirbeln – Norkitten – Nordenburg – Angerburg.
26th. Damerau – Wehlau – Allenburg – Gerdauen –Barten–Rosenthal.

The Main Body of the IInd Corps moved up on the left of the Army on the 25th, and the 56th Division passed into the XXth Corps to replace the casualties suffered in the 28th and 29th Divisions at Gumbinnen.

The Army now consisted of 4 Corps and 5½ Cavalry Divisions.

With the occupation of the Insterburg–Angerburg line, Instruction No. 1 had been carried out. It had taken nine days to complete the task and cover the 38 miles from the Frontier.

On the 26th a fresh task was communicated to the First Army, the Chief of Staff of the N.W. Front wiring to Rennenkampf :—

"The Commander-in-Chief has ordered the further objectives of the First Army as follows :—

(i) The blockading of Königsberg with a portion of its forces, approximating to two Corps, until relieved by Reserve Divisions.

(ii) The pursuit, with the remainder of its forces, of those enemy troops which do not take refuge in Königsberg, and which may be supposed to be withdrawing to the Vistula.

The combined operations of the First and Second Armies must aim at pressing towards the sea those Germans who are retreating to the Vistula and preventing them from reaching that river."

That this order was not issued until the 26th shows how little Jilinsky's Staff was in touch with administrative matters, and how little they appreciated the factor of time. The First Army was now given two distinct tasks, necessitating the splitting of the Line of Communications. The longest possible warning should have been given to enable a change to be made in the rear services. There seems to be no reason why this order should not have been given as soon as the strategic effect of Gumbinnen had been realised.

Jilinsky's order merits careful study. The blockading of Königsberg is placed first, and is evidently considered the more important task. From the wording of the order, "The pursuit of those enemy troops which do not take refuge in Königsberg," it is evident that he thought that a large proportion of the German forces would shut themselves up in the fortress, a view which is strengthened by the fact that two Corps are detailed for the blockading operations.

Jilinsky had made a fresh plan, based on the false impression that the Germans were retreating in disorder. He intended to hold back the First Army, while pressing forward the Second Army towards the North. That the Second Army would be in any danger from the disorganised Germans never entered into his calculations. This Northward move of the Second Army was intended to bring it closer to the First Army, but as we know, Jilinsky had failed to exercise proper control over Samsonoff, who had disobeyed his orders from the outset, persisting in deflecting his Advance farther and farther to the West, and away from the First Army.

Upon receipt of Jilinsky's orders, Rennenkampf issued Instruction No. 2, which ran as follows :—

Instruction No. 2.

The following dispositions will be carried out :—

(i) Königsberg will be closely blockaded. In order to shorten the line of blockade, the enemy will be driven back to the line of the R. Deime.

(ii) In view of the fact that the line of the R. Deime may be expected to be strongly held, preparations will be made for the forcing of that line in co-operation with a movement turning the right flank of the position.

(iii) When the line of the R. Deime has been occupied by the two Right Flank Corps and strengthened, the two Left Flank Corps will continue the Advance.

(iv) The Advance against Königsberg will commence with the forcing of the R. Deime.

(v) If the enemy assumes the offensive he will be driven back.

This deliberate action on the part of the First Army was evidently in accordance with Jilinsky's plan to drive the Germans towards the sea, for no comment was made upon Rennenkampf's dispositions.

The Russian decision to blockade Königsberg raises an interesting question as to the proper method of dealing with enemy fortresses during an advance.

The geographical situations of Antwerp and Königsberg, with reference to the Lines of Advance of the two great Invasions in Belgium and East Prussia, are very similar, yet they were dealt with in different ways. The Germans observed Antwerp, while the Russians decided to blockade Königsberg.

Both fortresses lay close to the right flank of the Invader's Line of Advance. The Belgian fortress was infinitely stronger than the East Prussian one, and a larger body of field troops might be expected to take refuge within it, owing to the probable plan of action attributed to the Belgian Field Army. Both fortresses could be reinforced or evacuated by sea, Antwerp *viâ* Ostend, and Königsberg *viâ* Pillau. Neither could be completely isolated. Both were of large extent, in that their subsidiary field works enclosed a considerable area of ground, within which it was possible to manœuvre. In both cases, sorties might be expected against the Lines of Communication, unless some action was taken.

The Russian advance had been hastily improvised, and no previous preparations for dealing with Königsberg had been made. The Reserve Formations were far behind the Main advance, owing to their unpreparedness and the insufficiency of the Lines of Communication to maintain them as well as the First-Line troops. Whether Königsberg was to be observed or blockaded, the use of First-Line troops at this moment was essential. There was no information as to how many German troops had retired into Königsberg, and Jilinsky, holding the opinion that a large proportion of the Eighth German Army would retire in this direction, detailed two Corps for the blockading, thereby making a certainty of superiority.

The German advance had been carefully thought out and elaborated in every detail. A Reserve Corps was moved up immediately behind the First-Line troops and placed in observation of Antwerp. The Main advance was not delayed for an hour. That the Germans ran some risk in observing the fortress with so small a force there can be no question, but in war risks must be run to achieve great results, and the Germans were justified by the results they obtained. The Garrison of Antwerp consisted of some 70,000 men, while the Belgian Army numbered 80,000 when it took refuge within the fortress. The two Reserve Divisions employed by the Germans numbered some 35,000 men only, and although the Belgians made two sorties, the Observing Force was maintained in position by reinforcing it temporarily by other troops passing through to the Front.

The lure of a fortress has often been fatal to a Commander, both in attack and defence. Sieges and blockades of Fortresses such as Antwerp and Königsberg, situated away from the decisive point, are subsidiary operations and should receive only limited attention until the decision has been obtained. The Germans threw nothing in the way of mobile troops into Königsberg, but the Russians allowed themselves to deflect two First-Line Corps to deal with it, to the detriment of the main operations farther West. Although the Belgian Field Army had withdrawn into Antwerp, the Germans only detailed a Second-Line Corps to observe it.

The Fortresses of Liége, Namur, and Maubeuge lay along the direct route of the main German advance; special forces were therefore detailed to isolate them, while the main German Forces passed them by on either side. But this was not enough, for the fortresses blocked the roads and railways, without which the German advance could not be maintained. Time was all-important, and therefore all available siege material was detailed to reduce the fortifications by hurricane bombardment, a method which met with complete success. As regards the little Fort Boyen, blocking the Lötzen Gap, the situation was only slightly different. The possession of the Gap was essential to the Russians, both for their future advance and to prevent any operations against their Communications. The Fort could not be overrun and isolated as the Germans did with the Belgian Fortresses, since the Lakes prevented such a manœuvre. The only method of reducing the Fort was by bombardment, and it would have paid the Russians to detail a few guns of large calibre to deal with the place, sending them afterwards to the Austrian Front. Here again the Russians failed through having made no previous preparations, and the failure to reduce Lötzen was to prove a thorn in their sides during subsequent operations.

Jilinsky had specially excluded any attack against Lötzen, stating that if Fort Boyen was to be attacked, fresh orders would be issued. No such orders were issued, and Rennenkampf began to feel the effect of this little fort remaining in German hands. The energetic Commandant, Colonel Busse, had carried out several sorties, and Rennenkampf wished to have done with him.

The detachments left behind by the IInd Corps were situated as follows :—

(i) The 170th Regt., with four guns and one Sotnia at Widminnen, opposite Lötzen.

(ii) The 169th Regt., with eight guns and four Sotnias at Arys, with a Battalion from the Regt. at Lyck.

On the 27th, on an order from Rennenkampf, the Commander of the 170th Regt. sent forward a Staff Officer with a trumpeter and an orderly to summon the German Commander to surrender. The message delivered concluded with the words, "You are given four hours in which to decide. If the fort is not surrendered within that time, not one stone will be left upon another."

The party was fired on by the Germans outposts, and the Staff Officer and his trumpeter wounded and taken prisoner.

Thus ended this inglorious attempt to frighten the Germans into surrender. No further attempts were made to gain possession of Lötzen.

During the 27th the reshuffling took place. The XXth Corps and the IIIrd Corps, with one Cavalry Division and Oranofski's Independent Cavalry Brigade, began to close up to the R. Deime.

The two Left Flank Corps moved as follows :—

IVth Corps. Friedland–Allenburg.
IInd Corps. Gerdauen–Rastenburg.

The IInd Corps still maintained its detachments to the East of the Masurian Lakes.

The Cavalry Corps, now three Divisions, was ordered to reconnoitre in front of the Army between the R. Pregel and the line Darkehmen–Gerdauen–Bartenstein.

Gourko's Cavalry Division was to reconnoitre South of the Cavalry Corps, being given the further task of ascertaining what the enemy was doing on the Western side of the Lakes near Lötzen, and of protecting the left flank of the Army during its Advance.

Here for the first time does one find the Cavalry Corps being sent forward in advance of the Army, but there are no instructions as to what they are to do—they are merely told to reconnoitre. Gourko is given a multiplicity of tasks, general protective duties with reconnaissances in varied directions.

It was late on the 27th that Jilinsky wired to Rennenkampf :—

> " Units from your front are being transferred to the front of the Second Army by rail, and are attacking strongly at Bischofsburg, Gilgenburg, and Soldau. Allenstein has been occupied by us. Co-operate with the Second Army by moving your left flank as far forward as possible towards Bartenstein, and your Cavalry to Bischofsburg. The VIth Corps has been ordered from Sczepanken to Passenheim."

Jilinsky still had a picture of the Second Army advancing Northwards, but no longer considered that Army capable of dealing with the situation alone, and was therefore moving forward the left flank of the First Army into the area in which he expected to find the Second Army within the next few days. The full extent of the retreat of the VIth Corps was not known to him, nor did he have any information as to the rout of the I^st^ Corps on the left of the Second Army. Jilinsky's order displays no uneasiness about the position of the Second Army. Had he had any inkling of the real situation, he must have directed the First Army much farther to the South, and would have ordered it to move forward by forced marches.

By the evening of the 28th, Rennenkampf had succeeded in moving three Corps as follows :—

IIIrd Corps. Friedland.
IVth Corps. Georgenau–Schönbruch–Klingenberg.
IInd Corps. Schippenbeil–Rössel.

By the evening of the 29th :—

IVth Corps. P. Eylau–Bartenstein.
IInd Corps. Gallingen–Bischofstein.
Cavalry Corps. Landsberg.
Gourko's Cavalry Division. Heinrichsdorf.

No contact with the enemy was gained, and there was no news of the Second Army.

2.—Rennenkampf's Withdrawal.

On the 29th, Rennenkampf received two telegrams from the Chief of Staff, N.W. Front :—

At 7 p.m.

"In view of the heavy fighting in the Second Army, the C.-in-C. orders you to move two Corps to its support. The Cavalry should be sent forward in the general direction of Allenstein."

At 11 P.M.

"The Second Army has withdrawn, therefore the C.-in-C. orders the forward movement of the two Corps to cease."

At this moment Jilinsky made an attempt to withdraw the IInd Corps from its position, with a view to sending it round by train from Suwalki to Bialystok and Ostrolenka, and thence to the area of the Second Army. He was dissuaded from this when he realised that such an operation—a march of 90 miles, followed by a railway journey of 160 miles—would take weeks to execute. That he should have conceived such a project is sufficient evidence of his lack of ability.

The positions occupied by the Corps of the First Army on the succeeding days were as follows:—

On the evening of the 30th:—

IIIrd Corps.	Friedland.
IVth Corps.	Georgenau–Schönau–Dietrichsdorf.
IInd Corps.	Paaris–Rastenburg.

On the evening of the 31st:—

XXth Corps.	R. Deime. North of the R. Pregel.
IIIrd Corps.	Wehlau.
IVth Corps.	R. Omet.
IInd Corps.	Between Lakes Nordenburg and Mauer.

Corps were therefore being withdrawn, but the Cavalry had been sent forward in one last attempt to help the Second Army.

Upon receipt of the order to send forward the Cavalry in the direction of Allenstein, Khan Nachivanski and Gourko were both directed to move off at once. The counter-order, received by Rennenkampf at 11 P.M., could not be passed on in time to stop them. General Gourko has given an account of his raid.

He was at Heinrichsdorf when the order arrived, prepared to give his Division a day's rest. His orders were "To find out and get evidence of the condition and whereabouts of the Second Army, news of which has been absent for two days." He was also told that Khan Nachivanski with three Divisions would be operating to the North of him. There were no orders, however, for the co-ordination of the two forces.

Gourko's many duties as to the protection of the left flank and reconnaissance towards Lötzen had tied him considerably to the rear, so that he knew little of what was going on in front of him. He moved off at 12 midnight on the night of the 29th/30th. Seeburg was passed to the South, after a brush with some German Railway Guards, and Allenstein sighted by 11 A.M. the next morning. The railway line outside the town was blown up as a precaution, the noise of the explosion soon bringing strong German Infantry forces out against him. A patrol arrived telling him that Khan Nachivanski was coming down from the North, but as there were no signs of him by 3 p.m., and nothing could be seen of the Second Army, Gourko decided to break off the fight and start on the return journey. By 8 P.M. the Division had covered 16 miles, and a halt was made till 11 P.M. Heinrichsdorf was reached once more at 6 A.M. on the 31st, after some fighting with German motor patrols. It was then found that the Russian line had withdrawn, and the march was continued to Kröligheim, which was reached at 6 P.M. Thus about 80 miles in 42 hours were covered on this useless enterprise. Both horses and men were thoroughly exhausted.

Khan Nachivanski had drifted off to Wormditt, where he had skirmishes with German Infantry, suffering some casualties, but making no effort to co-operate with Gourko.

By the 1st September the Russian Cavalry were located as follows :—

2nd Guard Cavalry Division	South of Wehlau.
Oranofski's Cavalry Brigade	On the R. Deime.
Khan Nachivanski (3 Divisions)	Landsberg,
Gourko's Division . .	Kröligheim.

On the R. Deime the Russians had made little progress, their attempts at seizing Tapiau, at the junction of the Rivers Pregel and Deime, being easily defeated by the Königsberg Landwehr.

On the 30th August Jilinsky wired to Rennenkampf :—

"General Samsonoff has suffered a complete defeat,

and the enemy has now complete liberty to turn against you. Keep this in view. You must take every measure to interrupt the railway lines by which the enemy may move troops against you. Pay attention that the enemy does not operate against you through Lötzen."

Jilinsky does not appear, however, to have given up the idea of blockading Königsberg, for on the 1st September his Staff was ordering the formation of the XXVIth Corps out of the 53rd and 56th Reserve Divisions for this purpose.

On the same day that Jilinsky conveyed the information of Samsonoff's disaster to Rennenkampf, a telephonic conversation took place between Generals Oranofski and Mileant, the Chiefs of Staff of the N.W. Front and the First Army respectively. It gives a picture of the conditions prevailing in the First Army at the time. Mileant suggested that the left flank of the Army should be withdrawn to Goldap, and that the Army should be concentrated on the general line of Insterburg. He stated that the Cavalry were exhausted with raiding, and considered that they should be withdrawn behind the line of Corps, in order to refit and get some rest. Oranofski agreed. When Mileant referred his project to his Chief, the latter refused to withdraw, saying that it would be bad for *moral*, and insisted on the Cavalry raiding being continued.

On the 1st September the Germans began to press the IIIrd Corps on the front Friedrichsdorf–Friedland, and news began to come in that a general move against the First Army was commencing.

On the 2nd September Rennenkampf issued Order No. 6, which ran as follows :—

Order No. 6.

The following line will be taken up and prepared for defence :—

XXth Corps.
1st Independent Cavalry Brigade—Line of the R. Deime.
2nd Guard Cavalry Division—North of the R. Pregel.

IIIrd Corps.
1st Guard Cavalry Division—Line of the Rivers Alle and Omet. Wehlau to Grünheim.

IVth Corps.
Khan Nachivanski (2 Divisions)—Grünheim–Nordenburg.

IInd Corps.
Between Lakes Nordenburg and Mauer, with a screen facing Lötzen from the East.

Gourko's Cavalry Division.
Angerburg. Thence to Goldap for reconnaissance on the left flank.

On the same day, after receiving Rennenkampf's proposed dispositions, Jilinsky issued the following :—

(i) The First Army will maintain its present position at all costs against the superior forces of the enemy which may be brought against it.
The Area between the lines Insterburg–Angerburg and Wylkowyszki–Suwalki, North of the Masurian Lakes, will be defended with obstinacy.

(ii) The Second Army will cover the approaches to the lower Narew and the area Augustow–Bialystok–Sokolka, taking up positions in the area Lomja – Przasnysz – Ciechanow–Mlawa, with Advanced Guards on the roads to Myszyniec, Chorzele, and Mlawa. This Army will be reinforced by the XXIInd, IIIrd Siberian, and Ist Turkestan Corps, to which are allotted the following areas :—

XXIInd Corps. Lyck–Arys–Johannisburg.
IIIrd Siberian Corps. Szczuczyn–Grajewo –Osowiec.
Ist Turkestan Corps. South of Lomja.

Up to the 4th September Jilinsky seems to have thought that a German advance against the First Army was certain, and his intention was to await it. But on this date the Germans under Von der Goltz with 2½ Divisions advanced on Mlawa, driving off the Russian Cavalry and seizing the town. From the moment the news of this advance arrived, the idea of an offensive became uppermost in his mind. The Second Army was not ready, and any advance by them must be carried out over the desert area North of Ostrolenka, so he turned to the First Army, which he had allowed to withdraw, and which was now in process of taking up a defensive position.

On the evening of the 4th he wired to Rennenkampf :—

"The undertaking of an offensive in East Prussia appears to be essential, not only on account of the general situation, but on account of the successes achieved in Galicia. I have indications that the Germans are advancing against Mlawa. The left flank Corps of the Second Army, which are concentrating in the area Przasnysz–Ostrolenka, will oppose this advance. In order to reinforce you, I propose to place under your Command the Corps which are now concentrating in the following places :—

IIIrd Siberian Corps. Lomja–Osowiec–Grajewo.
XXIInd Corps. Lyck.

If you consider an offensive possible by a gradual move to your left flank of the two Corps mentioned above, then I authorise you to advance in the direction of Bischofstein–Rastenburg, protecting yourself sufficiently against the direction of Königsberg. In view of the appearance of German Forces near Rhein and Johannisburg, these Corps will move as follows :—

XXIIrd Corps. Arys–Rhein.
IIIrd Siberian Corps. Johannisburg.

They will thus come up gradually behind the left flank of the First Army."

A few hours after this telegram to Rennenkampf, Jilinsky issued Instruction No. 4.

Instruction No. 4.

1. The leading portions of the XXIInd and IIIrd Siberian Corps have been directed towards Lyck. With the Ist Turkestan Corps they will form the Tenth Army under General Radkevich.

2. Boundaries between Armies will be as follows :—

   First and Tenth. Lake Löwentin–Widminnen–Marggrabowa.

   Tenth and Second. Rudczanny–Nowogrod–Sniadowo–Czyzew.

3. The Special Detachment of General Olkofski will cover Warsaw.

   Composition :—

   Caucasian Cavalry Division.
   5th Cavalry Division.
   Brigades of the 1st Turkestan Division.
   Guard Cossack Cavalry Brigade.
   Astrakhan Cossack Cavalry Regt.
   XXVIIth Corps (59th and 77th Divisions).

4. A General Advance into East Prussia is intended for the 14th September :—

   (i) Second Army.

   On the line Myszyniec–Chorzele, in co-operation with the Tenth Army.

   (ii) Tenth Army.

   XXIInd Corps. Between Lakes Löwentin and Spirding, co-operating with the First Army in the capture of Lötzen, by advancing against the place from the West through Rhein.

   IIIrd Siberian Corps. In the direction of Johannisburg and Rudczanny.

5. The First Army will stand fast during the concentration of the XXIInd Corp North of Arys and Lyck.

The positions of the troops of the N.W. Front on the night 5th/6th September were as follows :—

First Army.

XXth Corps. H.Q. Damerau.

28th Division. Paddeim–Gr. Schmerburg–Rudlacken.

29th Division. Taplacken. In Reserve.

XXVIth Corps.

56th Division.
53rd Division. } Rudlacken–R. Pregel.

IIIrd Corps. H.Q. Skungirren.

25th Division. Wehlau–Allenburg.

27th Division. Allenburg–Grünheim.

IVth Corps. H.Q. Georgenfelde.

30th Division.
40th Division. } Gneisenau – Gerdauen – Nordenburg.

IInd Corps. H.Q. Benkheim.

26th Division. Between Lakes Nordenburg and Mauer.

43rd Division. Facing Lötzen from the East.

72nd Division. Darkehmen.

54th Division. Insterburg.

1st Independent Cavalry Brigade. Grossbaum.

1st and 2nd Guard Cavalry Divisions. Rautzisken.

2nd and 3rd Cavalry Divisions. Hochlindenberg.

1st Cavalry Division. Buddern.

Tenth Army.

1st and 3rd Finnish Rifle Regts. Grajewo.

A Brigade of 8th Siberian Rifle Division. Augustow.

Portions of the 4th Finnish Rifle Regt. Kolno.

Portions of 1st Siberian Rifle Division. Kolno.

Portions of Ist Turkestan Corps. Lomja.

Second Army.

VIth Corps.

16th Division. Kadzidlo.

4th Division. Chudek–Jegliowjetz–Dylewo.

XXIIIrd Corps. Przasnysz.
Ist Corps.
  22nd Division. Stupsk.
  24th Division. Konopki.
  1st Rifle Brigade. Wola.
XXVIIth Corps.
  59th Division. Warsaw.
  77th Division. Warsaw.
XIIIth and XVth Corps débris. Zambrowo.
4th Cavalry Division. Myszyniec.
6th Cavalry Division. Przasnysz.
15th Cavalry Division. Chorzele.

3. Comments upon the Russian Movements after Tannenberg.

The disaster to the Second Army had come as a crushing surprise to Jilinsky. He had urged forward this Army, believing it to be about to intercept a disorganised and defeated enemy. Underrating his opponent, he had neglected to bring all his available forces to bear at the decisive point, and had directed the First Army to undertake a subsidiary operation in the blockade of Königsberg. The German blow came too quickly for him to be able to rectify matters before the Second Army had been defeated. Jilinsky's first thought was fear that the First Army would suffer the same fate as the Second, and he ordered an immediate cessation of the advance. If we examine the situations in East and West at the high-water marks of the two invasions, we find a striking difference in the strategic conceptions of the two High Commands. Both Germans and Russians had penetrated deeply into enemy territory, and their advances had been brought to a standstill for various reasons. The Germans had sustained a defeat on the Marne, the Russians had suffered a disaster at Tannenberg. Both for the moment had lost the initiative. Time was required to reorganise and bring up reserves. The Germans at once began to stablise the situation by fighting. The IXth and IXth Reserve Corps, on the 17th September, attacked and brought the French to a stand on the line Elincourt–Bailly–Carlepont. Again, on the 22nd, the IInd and XVIIIth Corps attacked the French in the region of Noyon. On the other hand, not only did the Russians not attack, but they withdrew a considerable distance. There could have been no immediate danger of a German attack against the First Army, for their forces must have been in some disorder as a result of the battle they had fought. The Russians allowed any chance there was of catching the Germans at a disadvantage to slip. Against a mobile enemy such as the Germans were, time could not be gained by a passive withdrawal, and Jilinsky should have ordered the First Army to manœuvre and fight, and despite the disadvantages of the geographical situation, to maintain as advanced a position as possible.

Six precious days elapsed before Jilinsky began to think of the possibility of an offensive, by which time the First Army had withdrawn still farther, and was settling down to the passive occupation of a defensive position. Scheme after scheme for another great concentric advance poured out in the form of telegrams to the Armies. The three fresh Corps coming up were allocated to the Second and

First Armies in turn, and were finally embodied in a new Tenth Army. On the 4th September Jilinsky issued orders for a general advance on the 14th, without much thought as to what the enemy would do in the 16 days since their victory at Tannenberg. He was dealing with phantom forces, as far as the Second and Tenth Armies were concerned, a fact which once more shows the administrative weakness of the Russian Command.

Rennenkampf was about to occupy a defensive position from the sea near Labiau to Angerburg at the Northern end of the Masurian Lakes. He issued his orders for the taking up of the position on the 2nd September, and he might therefore be sure of sufficient time to make himself secure. The position he had selected was of great strategical importance, in that it blocked the Insterburg Gap, and the line of the Rivers Deime, Alle, and Omet was naturally a strong one. His right flank was secure against the sea and the difficult country North of the R. Pregel. Here, however, the advantages of the position ended. The left flank could be turned by a German advance through the Lötzen Gap or round the Southern end of the Lakes, and the extent of the line was too great for the forces available. From the sea to Angerburg is 60 miles, and Rennenkampf had only 12 Divisions at his disposal, four of which were Second-Line and inferior to the others. The *moral* of the first Army also had been affected by the disaster to the Second Army, and its own losses sustained at Gumbinnen.

But it was in the method which he adopted in

occupying the position that Rennenkampf showed his inferiority. He proposed to carry out a purely passive occupation without any preliminary manœuvring. When the enemy has liberty of manœuvre, the passive occupation of a position can rarely be justified. He disposed his troops in a linear defence, without the least knowledge of where the enemy was likely to deploy his main strength, and all power of manœuvre was lost. Judging from his dispositions, he appears to have thought that the Germans would advance through Königsberg, for he placed four of his 12 Divisions on the R. Deime, on a front of 18 miles, while keeping seven Divisions between the R. Pregel and Angerburg, on a front of 42 miles. One Division was maintained as a defensive flank against Lötzen, to the East of the Lakes. An advance through Königsberg was the nearest route for the Germans to reach Rennenkampf's main Line of Communications through Insterburg–Gumbinnen–Kovno, which was essential to the maintenance of his Army, and the Germans adroitly induced him to believe that an attack was coming this way, by sending billeting officers to Königsberg, and spreading reports that troops from the West were about to arrive there.

If we accept the position chosen by Rennenkampf as the one which was vital to the general situation, then he should have disposed a portion of his forces well in advance of this position, with orders to manœuvre and develop the enemy's advance, and obtain the information without which the position could not be finally occupied with a proper economy of force. The remainder of his forces should have been detailed to reconnoitre the position and prepare it as far as time permitted.

It is therefore suggested that, as soon as Rennenkampf realised that he would probably be thrown on the defensive, he should have issued an Instruction upon the lines given below. This Instruction might have been issued on the night of the 30th or 31st of August.

4. Suggested Instruction to be issued by Rennenkampf.

First Army Instruction No. 4.

Gerdauen,
11 p.m., *30th August.*

1. The Second Army has been driven back from East Prussia after heavy fighting in the area of Neidenburg. The enemy is now free to direct his main strength against the First Army. The First Army will therefore assume a temporary defensive with a view to gaining time for the concentration of fresh Reserves and for the reorganisation of the Second Army.
2. A Defensive Position from Labiau to Angerburg, along the course of the Rivers Deime, Alle, and Omet, will be reconnoitred and prepared for defence.

   The First Army will withdraw fighting to this position if compelled to do so by superior force.
3. The IVth Corps and the IInd Corps (26th Division and the 27th Division of the IIIrd Corps) with

the Army Cavalry will maintain their positions in front of the defensive line as follows :—

IVth Corps. Pr. Eylau–Bartenstein.
IInd Corps. Bischofstein–Rössel.

Objective : to reconnoitre and delay the enemy's advance.

Khan Nachivanski (3 Divisions). Landsberg.

Objective : to ascertain what movement is taking place by road and rail in the area North of the line Heilsberg–Wormditt towards Königsberg. All railways running Northward will be interrupted.

General Gourko (2 Divisions). Rothfliess.

Objective : to ascertain what movement is taking place by road and rail—

(i) From the Allenstein–Insterburg railway line Northwards to the line Heilsberg–Wormditt.
(ii) From the West to the Lötzen Gap.

Corps Commanders will support the Cavalry in front of them if necessary, seizing all opportunities for offensive action without becoming seriously engaged. The line of withdrawal for Corps will depend upon the movements of the enemy, but will probably be as follows :—

IVth Corps. R. Alle Section.
IInd Corps. R. Omet Section.

4. The 43rd Division (IInd Corps) with the 72nd Division, under the Commander of the 43rd

Division, will drive in the German bridge-head to the East of the Lötzen Gap. A position will be reconnoitred and prepared with a view to preventing the enemy from debouching from the Gap to the East and to confining him to the West of the line of the Lakes. Fort Boyen will be kept under Artillery fire. Reconnaissances will be carried out towards Arys and Lyck with a view to keeping touch with the fresh Russian forces assembling in that area, and preventing surprise should the enemy decide to advance round the Southern end of the line of Lakes.

5. The Defensive Position will be organised as follows :—

    (i) R. Deime Section.
        53rd and 56th Divisions (under the Senior Divisional Commander). From the sea to the R. Pregel inclusive.

    (ii) R. Alle Section.
        XXth Corps. From the R. Pregel exclusive to Grünheim inclusive.

    (iii) R. Omet Section.
        IIIrd Corps (25th Division and 54th Division). From Grünheim exclusive to Angerburg.

All sections to be prepared for occupation by 4 Divisions. All work to be completed by the 6th September.

6. Army Headquarters will open at Gerdauen at 6 A.M. on the 31st August.

5. The German Movements after Tannenberg.

Ludendorff had quickly decided what he intended to do after his success at Tannenberg. Before the battle was over, on the 29th August, he had commenced moving troops Northwards to meet the expected advance of the First Russian Army. This Northward movement had been interrupted temporarily by General Sirelius' attack against Neidenburg, but was quickly resumed.

He might have moved South into Poland, to the help of the hard-pressed Austrian Army, as he was besought to do by the Austrian General Staff, and he might have reached Warsaw without much difficulty. This would have been a considerable strategic gain, but he would have been compelled to leave a strong detachment in his rear to watch Rennenkampf, and he would have had to cross the desert area which had tried Samsonoff so greatly. Every moment was of value, for the Russians had great reserves in their distant garrisons, and these were known to be coming up apace. The old Divisions of the Eighth German Army had now received their first reinforcements, and were flushed with victory. Two fresh Corps had arrived from the West. East Prussia was calling for deliverance, and the call was too insistent to be ignored. Ludendorff, therefore, kept steadily to his true strategic objective—the destruction of his most menacing opponent, Rennenkampf.

He developed his plans with customary speed and thoroughness. In outline they were as follows :—

(i) Von der Goltz with 2½ Divisions (one Landwehr Division and the Vistula Garrisons) to check any Advance on the part of the Second Russian Army from Poland.

(ii) Von François with three Divisions (Ist Corps and 3rd Reserve Division) and a Cavalry Brigade to move South of the Masurian Lakes, with the primary object of turning Rennenkampf's left, but at the same time dealing with any Russian Reserves coming up from the Augustow–Grajewo area.

(iii) Von Mackensen with two Divisions (XVIIth Corps) and two Cavalry Divisions, less the Brigade with Von François, to move through the Lötzen Gap against Rennenkampf's left.

(iv) Hindenburg with eight Divisions (XXth, XIth, Ist Reserve, and Guard Reserve Corps) to move direct against the Insterburg Gap.

(v) Königsberg Garrison (one Division) to occupy the attention of the enemy on the R. Deime.

(vi) Lötzen Garrison (one Brigade) to improve the bridge-head to the East of the Masurian Lakes.

Ludendorff was apparently well aware of the Russian dispositions, owing to their reckless use of wireless. For employing eight Divisions for the direct attack, and five Divisions in the enveloping attack, with the prospect of these latter being deflected to deal with Russian reserves, he has been accused of not making a proper economy of force, his chief accuser being Von François. But it must be remembered that troops sent to the East of the Lakes would be out of supporting distance of those to the West, and Ludendorff did not wish to find himself in an inferiority should Rennenkampf decide to advance at a critical moment when the German Forces were divided. The Lötzen Gap was narrow, and a Corps and two Cavalry Divisions would take a considerable time to defile across it, therefore more troops could not be sent by this route. Finally, the Corps newly arrived from the West had been detrained towards the North, and time would have been required to pass them to the South.

6. The Right Flank Attack against Rennenkampf.

Von der Goltz attacked and took Mlawa on the 4th September. He then moved his troops gradually Eastwards, driving back the Advanced Troops of the Second Russian Army as they attempted to move into East Prussia from the South, and neutralising any action on their part throughout the succeeding operations.

By the night of the 5th September, Von François' forces were situated as follows :—

3rd Reserve Division } Friedrichshof.
1st Cavalry Brigade }
1st Division. Olschienen. Advanced Guard. Puppen.
2nd Division. Lehmanen. Advanced Guard. Powalczin.

The 3rd Reserve Division and the 1st Division were ordered to move South of Lake Spirding, while the 2nd Division passed through the Gap at Nikolaiken, the first objective of the force being Arys.

By the night of the 6th September, positions were as follows :—

3rd Reserve Division } Gr. Turoscheln.
1st Cavalry Brigade }
1st Division. Johannisburg.
2nd Division. Nikolaiken.

Nikolaiken was found unoccupied, and the few Finnish Riflemen in Johannisburg were driven off with loss.

The Fight at Bialla.

On the 7th September, Bialla was taken by the 3rd Reserve Division after a short bombardment. The Russian Forces consisted of three Finnish Rifle Regts. (six Battalions) with two Batteries of Artillery, reinforced later by another Finnish Rifle Regt. and two more Batteries. These eight Battalions and four Batteries stood no chance against the German superiority, and they were driven helter-skelter Eastwards, pursued by the German Cavalry. Some hundreds of Finns were captured.

The easy capture of Bialla enabled the Main body of the 1st Division to turn Northwards against Arys, confining their operations against Bialla to a threat from the West.

By the night of the 7th September, positions were as follows :—

3rd Reserve Division. Bialla.
1st Cavalry Brigade. Pursuing the Finns Eastwards.
1st Division. Immediately South of Arys.
2nd Division. The line Gurra–Buwelno. N.E. of Arys.

The Fight at Arys.

At dawn on the 8th September the 1st and 2nd German Divisions advanced against the town from South and West. The Russian positions were treated to a short bombardment before the German Infantry advanced to the attack, and by 10 A.M. the town was in their hands. The Russian Forces in Arys consisted of three Battalions of the 169th Regiment of the IInd Corps, three Battalions of Finnish Riflemen from the XXIInd Corps, two Batteries of Artillery, and four Sotnias of Cossacks. This Force was driven Eastwards in great disorder, leaving over 1,000 prisoners in the hands of the Germans. It is interesting to note that the 1st German Division advanced over their peace manœuvre ground, which lay immediately South of the town.

By the night of the 8th September, positions were as follows :—

3rd Reserve Division. Drygallen.
1st Cavalry Brigade. Klaussen.
1st Division. Ranten.
2nd Division. Ruhden.

Von François had lost no time in Arys, and had turned North against Rennenkampf's left flank.

The enveloping movement had gone well. The enemy's Advanced Troops of some 15 Battalions had been caught without proper Artillery support and had been heavily defeated. News was now accumulating that more Russian troops were in process of concentration at Lyck, and it was important to catch these also in the same state of unpreparedness if possible. Von François therefore ordered the 3rd Reserve Division with the Cavalry Brigade to advance straight on Lyck, while the Ist Corps continued due North. News from the Lötzen Gap was also encouraging. The 1st and 8th Saxon Cavalry Divisions were to be passed across there for operations against the left flank of the enemy, and were to come under the Command of Von François. Orders were sent to them to move direct on Goldap.

The 9th September was to be a brilliant day for Von François.

The Fight at Lyck (Sketch J).

The 3rd Reserve Division started early from Drygallen. The Divisional Regt. of Cavalry with the Advanced Guard (49th Reserve Inf. Regt.) soon reported that the enemy were holding some Heights on either side of Baitkowen, some six miles S.E. of Lyck. The Divisional Commander at once turned the Advanced Guard off to the North towards Osranken–Sdeden, with orders to turn the enemy's right flank. The Artillery was ordered to reconnoitre positions from the Fichtenberg to Height 160. The Main Body (the 34th Inf. Regt. of the 6th Reserve Inf. Bde. and the 5th Reserve Inf. Bde.) was ordered to move up behind the Andreaswalde. The Divisional Commander rode to a Height near Dom Rakowken, from whence a good view could be obtained.

At 7 A.M. it was reported that Monczen and Susczen were occupied by the enemy, and that strong forces could be seen in and around Neuendorf.

The Divisional Commander then issued the following order :—

" The Division will attack.

" The 34th Reserve Inf. Regt. (head of the Main Body) will pass forward under cover of the woods South of Baitkowen, deploying on the line Baitkowen –Schnepien.

" The Advanced Guard will seize Mostolten.

" The Main Body will move up into Reserve to Dom Rakowken.

" The Artillery will take up positions about Osranken and South of Baitkowen."

At 10.15 A.M. the news arrived that Thalussen was also occupied by the enemy.

The next order was as follows :—

" The 34th Reserve Inf. Regt. will advance on Neuendorf, with its right on the line Height 180–Cziernien. The 1st Cavalry Bde. will accompany this advance.

" The Advanced Guard will push forward in the general direction of Susczen–Chroscziellen, covering itself towards Thalussen."

At 11.25 A.M. the Advanced Guard was ordered on Thalussen–Monczen–Chroscziellen, with the object of outflanking the enemy's right.

At 1.10 P.M. the Main Body was ordered to support the 34th Reserve Inf. Regt. against Neuendorf. The fighting near Thalussen had now become very heavy, the enemy being found in the woods

South of Karlssorge, and the Advanced Guard was unable to find the enemy's right.

At 4.30 P.M. it was reported that the enemy was holding Bobern, and that the attack of the Main Body was held up by superior forces near Cziernien.

The Divisional Commander had ridden forward to Height 165, where he took stock of the position. His troops were held up on a front of 6½ miles, and he had been unable to turn either of the flanks of the enemy's position. No Reserves were left, and casualties had been heavy. The hostile Artillery fire was increasing in volume, and the guns had evidently been reinforced. Every moment might bring the Russians reinforcements. Tired as his men were, he considered the chances of a night-attack that night. He finally abandoned the idea in favour of a dawn-attack the next morning.

At 3.30 A.M. on the 10th, all stood ready for the attack. With the first few shells the enemy began to evacuate their positions, and by 10 A.M. the town was in the hands of the Germans. The enemy withdrew in great disorder, pursued by the Cavalry. Few prisoners were made, but the casualties on both sides were heavy. The Russian Forces had consisted of the remainder of the XXIInd Corps, about twelve Battalions, with the majority of its Artillery, consisting of nine Batteries. The opposing forces were therefore identically equal, 12 Battalions and 72 guns.

The marching of the 3rd Reserve Division is remarkable. Over execrable roads, with a sharp fight at Bialla on the 7th, they had covered 55 miles in 3 days (6th, 7th and 8th). They had then fought all day on the 9th, marching into Lyck early on the 10th. The boldness of the Commander of this isolated Division had been fully justified. He had handled his troops with great energy, employing all his reserves to prevent the Russians from taking part in the main action about to be fought on his left, and preparing to attack again on the second day when checked on the first. His tactical dispositions and orders are an example of what they should be in such a battle. It is interesting to contrast his bold action with that of the Commander of the VIth Russian Corps at Bischofsburg on the 26th August, the opening day of the battle of Tannenberg. Small detached forces, carrying out a protective rôle, depend upon their mobility and offensive spirit to achieve their purpose. The 3rd Reserve Division had fulfilled its rôle to a nicety.

It was after this defeat at Lyck that Jilinsky ordered the XXIInd Corps to concentrate at Augustow, and the IIIrd Siberian Corps to assemble at Grajewo. These two Corps had lost 8,000 men in the fighting of the 7th, 8th, and 9th, a fact of which Jilinsky must have been unaware, for he made no mention of their broken condition in informing Rennenkampf of their withdrawal. The left flank of the First Army was now exposed to the attack of the Ist and XVIIth German Corps with 2 Cavalry Divisions.

During the night of the 8th/9th September, Von François got into touch with the XVIIth Corps, which had been crossing at Lötzen. The Commandant of Fort Boyen had pushed out his bridgehead to the line Stasswinnen–Schedlisken–Kruglinnen–Soldahnen–Pietzarken, where he found himself in close touch with the Russians in superior force. He had received no Instructions as to the preparation of the Lötzen Gap for the passage of large forces, and nothing had been done to improve the communications. The Gap is only about a mile across at its widest point, and it is traversed by one good road and the track of the railway. The whole of the 8th had been employed in passing the XVIIth Corps across and in moving up to the line of the bridge-head. All efforts at debouching farther East had been driven back. The Corps had attacked at 5 A.M., 12 noon, and 8 P.M., without success. The two Cavalry Divisions were still to the West of the Gap, which was blocked with the rear services of the XVIIth Corps. They had therefore been unable to obey their orders to move on Goldap. The difficulties encountered at the Gap are examples of how the neglect of an administrative matter can seriously affect the tactical situation. The Russians were strongly entrenched between the various lakes, and no means could be found of turning their flanks. The original left flank had consisted of the 43rd Division of the IInd Corps, less the four Battalions of the 167th Regt. at Arys and Lyck. As the day wore on they had been reinforced by the 72nd Division from Darkehmen, and the 54th Division from Insterburg. The Russians were therefore superior in numbers to the XVIIth Corps attacking them.

The Fight at Soltmahnen (Sketch K).

Von François lost no time in his preparations.

At 3.30 A.M. on the 9th he launched his two Divisions :—

1st Division.	On Widminnen, between Lakes Gablick and Soltmahn, and thence on Regulowken, between Lakes Siewen and Soltmahn.
2nd Division.	On Kruglinnen, between Lakes Soltmahn and Siewen, and thence on Gansenstein.

The attack was almost a complete surprise. The enemy were found holding the intervals between Lakes Gablick, Soltmahn, Siewen, and Babken with only weak forces. A fierce bombardment of 20 minutes was sufficient to make them vacate their positions and retire in disorder. Von François' men pushed straight forward in the immediate rear of the Russians opposing the XVIIth Corps. The whole Russian line then broke in hopeless rout, leaving guns and waggons and all their impedimenta behind them. Over 5,000 prisoners and some 60 guns were taken by Von François. The two Regiments of Divisional Cavalry were sent in pursuit, but the two Cavalry Divisions did not arrive at Widminnen until the evening, when it was too late to take part in the action. The delay at the Gap may well have been unavoidable, but it might have been reduced considerably, had there been sufficient administrative forethought.

The enveloping movement had met with complete success. The credit lies with Von François. From Olschienen, on the morning of the 6th, his men had covered 77 miles in 4 days (6th, 7th, 8th, and 9th), with deployments and fighting on the 8th and 9th.

The Russian Main Army now lay open to the

advance of two German Corps with two Cavalry Divisions, and another Tannenberg appeared within reach of Hindenburg's Army.

7. The Main German Advance against Rennenkampf. (Map III.)

The main German Advance against the Insterburg Gap was being carried out with 8 Divisions against 7 Russian Divisions. The XIth and Guard Reserve Corps from the Western Front had been detrained at Allenstein and Elbing, and had commenced their forward march on the 4th and 3rd September respectively.

Ludendorff employed the 5th, 6th, and 7th in bringing his 4 Corps into line.

The positions occupied were as follows :—

Night of the 5th :—
XXth Corps. Rothfliess.
XIth Corps. Seeburg.
Ist Reserve Corps. Heilsberg.
Guard Reserve Corps. Mehlsack.

Night of the 6th :—
XXth Corps. Rössel.
XIth Corps. Bischofstein.
Ist Reserve Corps. Heilsberg.
Guard Reserve Corps. Pr. Eylau.

Night of the 7th :—
XXth Corps. Rastenburg.
XIth Corps. Glittehnen.
Ist Reserve Corps. Wöterkeim.
Guard Reserve Corps. Georgenau.

Ludendorff in his Memoirs states that on the 7th the Army was entrenching itself in front of the enemy's line Angerburg–Nordenburg–Gerdauen–Wehlau. This makes it appear as if he were waiting for his enveloping movement to take effect before he carried out a frontal attack against the Russian positions. But the situations given above have been verified from the archives of the German General Staff, and agree also with the Russian accounts. It was not till the 8th that the Germans gained their first contact with Rennenkampf. By that night, after some skirmishing, they had dug in Advanced Posts on the general line :—

> Fürstenau - Alt Jaglack - Sausgorken - Molthainen - Momehnen - Mehleden - Schönbaum–Gr. Wohnsdorf - Karschau - Gr. Engelau.

The advance had been carried out in no haste, the daily marches averaging 11-12 miles.

Rennenkampf had pushed forward no troops in front of his main position, and it was not till the 8th that he realised that no attack was coming through Königsberg. The threat against his left flank was growing serious, and to meet it he despatched the 29th Division of the XXth Corps, then at Taplacken in Reserve to the R. Deime front, to the extreme left flank at Goldap. The crow-fly distance between these two places being 45 miles, little could be expected from this Division for several days. The early commitment of his troops to a linear defence had begun to find Rennenkampf out.

On the 9th September the Germans delivered an attack along the whole front of the Insterburg Gap.

The XXth Corps on the right was heavily repulsed, and the remaining Corps made little progress. The Russians were very strongly entrenched, and the German Artillery had failed in consequence to have its usual annihilating effect.

The situation on his left, to the East of the Lakes, now decided Rennenkampf upon a general retirement. Orders were issued for this during the evening. The critical position of his Army seems to have galvanised Rennenkampf into action, for he ordered a counter-attack between Angerburg and Nordenburg against the XXth and XIth German Corps. This counter-attack was most gallantly executed by the 40th Division of the IVth and the 26th Division of the IInd Corps. It was completely successful in its purpose, especially in the case of the XXth Corps, which did not thereafter cross the Russian defensive line until the 11th.

It is difficult to trace the Russian movements during their withdrawal from the Insterburg Gap, but the sequence of events appears to have been as follows :—

The withdrawal commenced at nightfall on the 9th, and by 7 A.M. on the 10th the following positions had been reached :—

XXVIth Corps. Mehlauken.
28th Division (XXth Corps). Wirbeln-Norkitten.
IVth Corps. Kl. Potauren.
IInd Corps. Nordenburg.

There was considerable confusion from the commencement, owing to the orders not having been issued until a few hours before the withdrawal was timed to begin. Many units received no orders.

The IInd Corps, thanks to its counter-attack, succeeded in extricating itself without difficulty, and by the evening of the 10th the following positions had been reached :—

XXVIth Corps. Spannegeln.
28th Division (XXth Corps). Gr. Bubainen.
IIIrd Corps. Jodlauken.
IVth Corps. Kowarren.
26th Division (IInd Corps). Dombrowken.

The Russian Corps had covered 20 miles in the first 24 hours of the withdrawal. Only on the left of the IIIrd Corps had the Germans pressed their pursuit.

By the night of the 10th the German Main Advance was along a general line of :—

Salzbach-Barten-Abelischken-Gr. Guie-Ilmsdorf.

As we know, to the East of the Lakes the German enveloping movement had destroyed the Russian left flank on the 9th. The news of this success produced the following order from Ludendorff on the evening of that day :—

"The Ist Corps will cover the Advance of the XVIIth Corps by advancing to the line Lissen–Benkheim. The Line Benkheim–Paulswalde–Klinken is not to be passed by either Corps for the present, since a counter-offensive from Insterburg may be expected."

With the memory of the Russian counter-attack

in his mind, Ludendorff was becoming cautious. By indicating the line Paulswalde–Klinken for the XVIIth Corps, he was pulling in his right flank to fill up the gap made by the defeat of the XXth Corps. This caution on the part of Ludendorff has been put forward as the reason for his failure to achieve another Tannenberg, but we must remember that the casualties in the XXth Corps had been heavy, and that the XIth Corps was new to the country and the enemy. This successful counter-attack exemplifies the effect which may be produced by a tactical operation upon the strategical situation.

By the night of the 10th the German right had reached the following positions :—

Ist Corps. Benkheim.
XVIIth Corps. To the left of the Ist Corps.
1st Cavalry Division. South of Goldap.
8th Saxon Cavalry Division. Goldap.

The Russian Cavalry had taken no part in the fighting, being reported as too exhausted to manœuvre. By the evening of the 10th they were situated as follows :—

Khan Nachivanski (2 Divisions). Kowahlen.
2nd Guard Cavalry Division. Kleszowen.

Gourko's Cavalry Division had moved down past Goldap in front of the German Cavalry. At this critical moment Jilinsky had decided to transfer him to the Tenth Army. He took no further part in the fighting.

The 1st Guard Cavalry Division and Oranofski's Brigade were away in the North beyond the R. Pregel.

Rennenkampf's next order, issued on the night of the 10th, was as follows :—

" The withdrawal will be continued to the line :—

" Kraupischken–Gawaiten–Goldap.

" And further to the line :—

" Kussen–Trakehnen–Tollmingkehmen.

" Final line to be occupied by the night of the 11th.

" The IInd Corps, with the 72nd and 54th Divisions, will cover the withdrawal of the XXVIth, IIIrd and IVth Corps.

" The XXth Corps will protect the left flank by energetic action towards Goldap."

The 28th Division of the XXth Corps had been ordered across the rear of the withdrawal to join its sister Division on the left flank. When this order was issued is uncertain. Its execution must have added considerably to the confusion in the Russian rear.

By the night of the 11th, the Main German Advance reached the following positions :—

XXth Corps. Launingken.
XIth Corps. Kuinen.
Ist Reserve Corps. Insterburg.
Guard Reserve Corps. Norkitten.

The enemy was found to be moving steadily in front of the German line, and there seems to have been little attempt made to press him.

Ludendorff had evidently been reassured during the night of the 10th/11th, for his orders to his enveloping flank for the 11th are bolder. They ran as follows :—

> 1st and 8th Saxon Cavalry Divisions. Wirballen–Kovno Road.
> I[st] Corps. On Mariampol. Moving S.E. of Rominten Forest.
> XVIIth Corps. North of Rominten Forest on Wisztyniec.

The I[st] Corps reached Goldap by noon on the 11th, and was preparing to move South of the Forest when another counter-order arrived. Ludendorff's heart had failed him again.

Ludendorff's fresh order directed the I[st] Corps on Gawaiten and the XVIIth Corps on his left against Kleszowen. The XIth Corps had shown nervousness as to its right flank, in the absence of the XXth Corps, and had reported that a Russian attack was imminent.

During the afternoon of the 11th, therefore, the I[st] and XVIIth Corps closed in to the North. The Russians were discovered entrenched along the line Gawaiten–Kleszowen. Both Corps Commanders decided that it was too late to attack that night, and prepared for an attack at dawn the next day. Von François, with his eye ever on envelopment, disobeyed his orders in so far as to send his 1st Division to Tollmingkehmen, which was reached without opposition.

The 8th Saxon Cavalry Division had taken 1,000 prisoners in Goldap on the evening of the 10th, but had lost a good many officers in the street-fighting.

For the night of the 11th, the German Cavalry were situated as follows :—

1st Cavalry Division. Mierunsken.
8th Saxon Cavalry Division. Dubeningken.

The Russians had been marching day and night without cessation. Their movement had been impeded by the streams of transport and parks blocking all the roads. The situation can only be compared to the Austrian withdrawal from the Italian Front in 1918. The havoc which could be wrought by a modern air force under such conditions can well be imagined. The wretched Russians were at least spared such slaughter.

As far as can be ascertained the succeeding positions of the Russians were as follows :—

XXVIth Corps. Kussen. By dawn on the 12th.
IIIrd Corps. Trakehnen. By dawn on the 12th.
IVth Corps. On the R. Angerapp near Dinglauken. Night of 11/12th.
IInd Corps. Kariotkehmen–Pabbeln. By the night of the 11/12th.
XXth Corps. On the line Gawaiten–Kleszowen. In contact with Germans.

The condition of the Russian Army can now be imagined. The XXVIth Corps had covered 58 miles in 60 hours, and the IIIrd Corps considerably more. When it is remembered that the weather was extremely hot, and that all the rear services were blocking the roads, it is clear that the offensive power of the First Army must have sunk very low.

Rennenkampf appears to have broken down under

the strain of events, for he issued a final order to his Corps at 5.30 P.M. on the 12th :—

> IIIrd Corps. Wylkowyszki–Kovno.
> IVth Corps. South of the IIIrd Corps.
> XXVIth Corps. Sredniki.

Having issued this order, the Army Commander betook himself to Kovno with his Staff, imitating the example of Samsonoff in cutting off all communication with Jilinsky. The Commander of the N.W. Front, in reporting to the Supreme Command, said :—

" General Rennenkampf has thought more about the safety of his Staff than of directing the movements of his Army, which he has not in actual fact commanded for several days. He reports that he is moving personally to Wylkowyszki, and that he is withdrawing the IIIrd and XXVIth Corps Eastwards, leaving the remaining Corps of his Army to their fate. His Staff has gone with him. The behaviour of the Army Commander has made all direction of operations impossible. He has altered the position of his Headquarters four times in the last 24 hours, each time completely breaking off communication."

Supreme Headquarters did not appear to have much sympathy with Jilinsky, for they replied that it was quite in the nature of General Rennenkampf to wish to direct his troops personally, and that he had probably gone to Wylkowyszki for this purpose. The final injunction was, " Try and get into touch with him there."

On the 13th, Jilinsky received the news that Rennenkampf was in Kovno. In reporting this fact to Supreme Headquarters, he added: "This information appears so improbable that it requires verification."

Von François says that the night of the 11th/12th at Eighth German Army Headquarters was a stormy one. Orders were sharp and peremptory. Both the XIth and XXth Corps received angry injunctions to push forward, and he himself was accused of holding up the advance of the XVIIth Corps. Ludendorff had at last realised the true position of affairs in the Russian Army.

For the 12th, the Ist Corps was ordered on Pillupönen-Wylkowyszki, while the two Cavalry Divisions were ordered South of the Forest of Rominten.

The Fight at the Forest of Rominten (Sketch L).

Von François calls the fighting of the 12th "interesting." It was certainly full enough of interest to please the most fighting spirit.

The 1st Division had spent the night of the 11th/12th along the road from Tollmingkehmen to Gr. Rominten. It was ordered to skirt the Northern edge of the Forest to Mehlkehmen, and thence to Pillupönen.

The 2nd Division, which had been in contact with the Russians at Gawaiten, was ordered to turn about and move down the road through the centre of the Forest to Wisztyniec.

Both Divisions were ordered to march with strong

Advanced and Rear Guards, and to be ready to deploy in any direction.

At 11 A.M., the Rear Guard of the 1st Division came into action against Russian columns trying to break Southwards into the Forest at Oszeningken. The fight became so brisk that a portion of the Main Body had to turn back to the assistance of the Rear Guard. The Russians were finally driven off Northwards, with the loss of 1,000 prisoners. The view from the Northern edge of the Forest is extensive, and the German Artillery must have had magnificent targets against the Russians advancing and retiring over the open fields. The Russians probably belonged to the IVth Corps making for the line Pillupönen–Göritten.

The Advanced Guard of the 1st Division reached the neighbourhood of Pillupönen without opposition. During the afternoon, strong Russian attacks developed from the North, attempting to break through. The remainder of the Main Body came up to the support of the Advanced Guard, and the attacks were driven off with heavy loss to the Russians.

Numerous small bodies of all arms were seen breaking in and out of the Forest, and in the plain towards Enzuhnen long bodies of disorganised troops and transport were moving N.E. across the fields. At this moment the two German Cavalry Divisions might have reaped a rich harvest, but they were now far to the South of the Forest.

The 2nd Division succeeded in disengaging from Gawaiten without any hindrance from the Russians. These were part of the XXth Russian Corps, which had been ordered to protect the withdrawal by

energetic action towards Goldap. After having allowed the Germans to turn their backs to them, they also moved off to the N.E. The 2nd Division then passed the Emperor's chalet to Nassawen and Mehlkehmen. The road through the Forest is good, but the Forest itself is almost impenetrable, except by the rides cut for the herding of the game. The Forest proved full of small bodies of Russian Infantry and Cavalry, and the effect of the German march down the centre was to drive these small parties out in all directions. Von François's Headquarters Column was cut up by Cossacks and all his baggage captured, but was rescued once more by a troop of Divisional Cavalry.

Von François himself seems to have spent a happy day moving about with a small Cavalry escort between his various detachments, encouraging them to hunt out the Russians.

By the night of the 12th, the Germans occupied the following positions :—

1st Cavalry Division. Wizajny.
8th Saxon Cavalry Division. Krolowe.
Ist Corps. Nassawen–Mehlkehmen–Pillupönen–Oszeningken.
XVIIth Corps. Darkehmen.
XXth Corps. Perkallen.
XIth Corps. Schröterlauken.
Ist Reserve Corps. Gr. Wannagupchen.
Guard Reserve Corps. Skattegirren.

Between 6 A.M. and 9 A.M. on the 13th, the Russians had managed to concentrate as follows :—

XXVIth Corps. Wladyslawow.
IIIrd Corps. Wylkowyszki.
IVth Corps. Gieystoryszki.

The débris of the XXth and IInd Corps also arrived at Wylkowyszki.

The Cavalry remained covered by the Infantry as follows :—

Khan Nachivanski. Kalwarja.
2nd Guard Cavalry Division. Pilwiszki.

A comparison of the German and Russian positions on the evening of the 12th, therefore, shows clearly that all chance of envelopment had disappeared.

8. Jilinsky's Final Efforts to save the First Army.

Cut off from communication with Rennenkampf, Jilinsky could have known little of what was going on. On the afternoon of the 12th, he managed to get in touch with the Commander of the IIIrd Corps, and ordered him to assume command of his own and the XXVIth Corps and advance at once in the direction of the enemy wherever he was, in order to save the rest of the Army. No attempt was made to carry out this order. On the same day, he spoke to a Staff Officer in Vilna as follows :—

"Try by all possible means to get the following message to General Rennenkampf. I have ordered the XXIInd Corps to march at once from Augustow to Marggrabowa. Moreover, five Battalions of the IInd Caucasian Corps will be to-day in Filipowo. To-morrow, the IIIrd Siberian Corps is concentrating at Grajewo for an advance to Lyck."

None of these advances were carried out in actual fact.

On the 12th, also, the Second Army was ordered to advance towards the East Prussian Frontier. Units of the Army did carry out certain movements towards Chorzele, but retired once more as soon as they came into contact with the German Cavalry.

On the 13th, Jilinsky received the following Instruction from Supreme Headquarters, signed by General Daniloff :—

"The Generalissimo has ordered the following :—

(i) In view of the conditions which you have explained as existing on the First Army front, it is useless to order the Tenth Army to advance, since it cannot now save the position. The Tenth Army will therefore cover the approaches to Grodno and Bialystok.

(ii) Corps of the Second Army will not engage the enemy in front of them, and will secure their freedom to withdraw to the Narew.

The main task of the Second and Tenth Armies is to prevent the exposure of portions of their forces to defeat in detail and to operate in close liaison, maintaining their hold over the Grodno–Bialystok area. The Cavalry will reconnoitre in front of the Armies and on their flanks in order to prevent surprise."

The Grand Duke wished to prevent any further

destruction of the Second and Tenth Armies by driblets, which had been Jilinsky's method of operating.

It is curious that an Instruction to employ the Cavalry in front of the Armies to prevent surprise should have been necessary in the Orders of a Supreme Headquarters of a First-Class Military Power.

On receipt of this Instruction, Jilinsky ordered the First Army to continue its withdrawal to the Middle Niemen, covering the crossings at Preny, Olita, and Merecz. He repeated the orders for the Second and Tenth Armies verbatim.

By this time, relations between Jilinsky and Supreme Headquarters were considerably strained. Rennenkampf also began to ignore Jilinsky altogether.

For the 14th, Rennenkampf ordered the First Army to withdraw to the line Szaki–Mariampol–Ludwinow. Orders for this movement did not reach their destinations, and there was much confusion. Rennenkampf placed the IIIrd and IVth Corps under the senior Corps Commander, but both Corps proceeded to move independently. The IIIrd Corps, moving from Wylkowyszki towards Szumsk, found the IVth Corps moving directly across its front from Gieystoryszki towards Kovno. The XXVIth Corps moved straight on to the Niemen at Sredniki. The débris of the IInd and XXth Corps followed the IIIrd Corps. The Germans did not press the retreating Russians anywhere.

On the night of the 14th, Rennenkampf wired direct to the Grand Duke, "All Corps have now broken off the battle." To this message he received the following reply :—

" With all my heart I thank you for your successful leadership. Convey my thanks to the heroic First Army for their efforts. For the future we place our faith in your energy and in God's help."

Rennenkampf reported also that the Corps of the First Army required only a short breathing-space to enable them to be ready for action once more. A few days later, however, Januschkevitch or one of the Staff of Supreme Headquarters visited Kovno and learnt the true state of affairs. As a result, on the 20th, Januschkevitch wrote to Rennenkampf as follows :—

" I wish to draw your attention to the unreliability of your report regarding the efficiency of your troops. The Generalissimo wishes you to know that he is obliged to depend upon written statements as regards the condition of the Armies, and only by this means can he draw a correct picture of the general state of affairs."

On the 15th, Jilinsky wired to Rennenkampf :—

" If the First Army is called upon to fight, its position does not appear safe, since both its flanks are exposed to a turning movement. Therefore, in order not to expose the Army to a dangerous blow, it is essential that you make all arrangements for a night-withdrawal. Units should cover 20–30 versts (14–19 miles) during the night."

By the night of the 15th, some sort of order was

being arrived at amongst Corps. Their actual positions appear to have been as follows :—

XXVIth Corps. On the left Bank of the R. Niemen near Sredniki.
IIIrd Corps. In the area Kozliszki–Grabowo.
IVth Corps. Mariampol.
IInd Corps. Débris assembling at Mariampol.
XXth Corps. Débris assembling at Ludwinow.
Khan Nachivanski. Sereje.
2nd Guard Cavalry Division. Kantalischki.

The final line reached extended from Sredniki–Kozliszki–Morgowo–Simno. They had thus reached the line from which they had started on their advance exactly a month previously.

On the 16th, Jilinsky issued his final Instruction :—

Instruction No. 11.

The enemy is concentrating his forces in the region of Suwalki, with strong detachments on the front Ortelsburg–Mlawa, and in rear of the retiring First Army. In view of the unsuccessful actions which have taken place on the latter front, the following dispositions will be carried out :—

(i) The First Army will withdraw to the Middle Niemen, covering the crossings between Kovno and Merecz.

(ii) The Second Army will concentrate in the area Sniadowo–Gonsewo–Wyszkow, covering the crossings of the Lower Narew from the directions of Lomja, Ostrolenka, and Rozan, ready to operate in close liaison with the Tenth Army.

(iii) The Tenth Army will cover the area Bialystok–Augustow, paying particular attention to the approaches to Grodno from Augustow and Sopockinie, and watching the Northern and Western exits at the Forest of Augustow.

(iv) The Warsaw Detachment will protect Warsaw.

(v) The Cavalry will reconnoitre on the fronts as follows:—

First Army. Sredniki–Sejny.
Tenth Army. Sejny–Bialla.
Second Army. Bialla–Mlawa.

On the 17th September, Jilinsky was replaced in his command by General Russky.

9. Comments upon Ludendorff's Manœuvre.

The conception of an envelopment of the enemy, aimed at from the very commencement of the operations, is clearly evident in Ludendorff's dispositions for the advance to the Masurian Lakes. It has already been pointed out how the fear of a counter-attack on the part of Rennenkampf caused Ludendorff to employ only five Divisions in his enveloping wing, while retaining eight with his frontal advance. The timing of the operation was very accurate, the culminating date of the long advance being the 9th September, curiously enough the same day as the Battle of the Marne in the West. On this date, the frontal advance commenced its holding operations and the enveloping wing dispersed the last remaining troops of the Russian left flank defence. The 10th and succeeding days of September seemed to promise the certainty of another Tannenberg for Ludendorff.

Rennenkampf did not wait for the events of the 10th, but ordered a general retirement to commence on the night of the 9th/10th. The quick realisation of his danger thus gave him a start of some hours. To increase this start Rennenkampf ordered a counter-attack on his left. This counter-attack was successful in checking completely one Corps and in delaying another considerably. The gap between the two German wings was thereby widened, and Ludendorff's orders for the 10th are cautious. He limited the advance of his enveloping wing, holding it ready to move Northwards to the support of the exposed right of the frontal advance.

For the 11th, Ludendorff's orders were bolder, but during the day he became apprehensive once more of a possible counter-attack, and again limited the advance of his right. Von François, the Corps Commander on the extreme right, as at Tannenberg, disobeyed his orders. But this time his disobedience extended only to the moving forward of one of his Divisions beyond the forbidden line and not to a bold push for envelopment.

By the night of the 11th, Ludendorff seems to have realised that the Russians were beyond the stage of counter-attacking, and his orders were sharp and peremptory, as if he had realised his mistake and was venting his spleen upon his subordinates. The two Divisions of Cavalry were sent off in a wide sweep to the South of the Forest of Rominten in a last hope of regaining a position from which to envelop the enemy.

The events of the 12th were to show that envelopment was then impossible. Ludendorff had gone from caution to the opposite extreme, and had directed his Cavalry beyond the power of co-operation with the enveloping wing. He was undoubtedly attempting to control too much. He was issuing orders to eight if not nine distinct units dispersed over a wide front. He had four Corps and the Königsberg garrison in his frontal advance and two Corps and two Cavalry Divisions in his enveloping wing. He would have done better to have placed the whole of the enveloping wing under one Commander, or if the personalities of Von Mackensen and Von François made this impossible, to create a smaller enveloping force by placing the two Cavalry Divisions under the Commander of the right Corps. If, on the 12th, envelopment was impossible, there was still a magnificent opportunity for mounted action against the demoralised Russians, the effect of which might have been very great.

It may be that aerial reconnaissance in the future will give a Commander a clearer idea of the state of demoralisation into which a retreating enemy has fallen, and that it will be possible to direct the pursuit to better advantage. Ludendorff seems to have failed to realise that the Russians were in disorder, and with the strategical situation less critical than it was at Tannenberg, he seems to have erred on the side of caution.

From the 13th onwards, the German pursuit gradually slackened. It was not Ludendorff's intention to pursue the Russians far into Russia. East Prussia had been freed for the moment and there was other work to do.

10. Results of the Battle of the Masurian Lakes.

The last of the Russian Forces had now been driven from East Prussia. Rennenkampf's troops had been exactly 28 days in German Territory. His losses during that period have been given officially by the Russians as follows :—

	Men.	*Guns.*
Stallupönen . . .	3,000	—
Gumbinnen . . .	17,000	—
Masurian Lakes . .	125,000	150
	145,000	150

As was the case with Tannenberg, two Corps, the XXth and IInd, suffered the greatest loss. But the whole of Rennenkampf's Army was unfit for active operations for a considerable time. Over half the transport had not succeeded in escaping, and the *moral* of the Army had sunk to a low ebb. The sending forward of the Tenth Russian Army in driblets had resulted in the loss of a complete Corps and in the demoralisation of many units in their first action.

11. Table of Russian and German Forces employed in East Prussia up to 13th September 1914.

(*a*) *Russian.*

Infantry	Ist, IInd, IIIrd, IVth, VIth, XIIIth, XVth, XXth, XXIInd, XXIIIrd	Corps.
	1st, 5th	Rifle Brigades.
	53rd, 54th, 56th, 57th	Divisions.
Cavalry	1st Guard, 2nd Guard, 1st, 2nd, 3rd, 4th, 6th, 15th	Cavalry Divisions.
	1st Independent Cavalry Brigade.	

A Total of:
- 410 Battalions.
- 232 Squadrons.
- 1,392 Guns.

(b) German.

Infantry	Ist	Corps.
	XIth	
	XVIIth	
	XXth	
	Guard Reserve	
	Ist Reserve	
	3rd Reserve	Divisions.
	1st Landwehr	
	2nd Landwehr	Brigade.
	Königsberg	Garrisons.
	Vistula	
	Lötzen	
Cavalry	1st	Cavalry Division.
	8th Saxon	

A Total of 224 Battalions.
128 Squadrons.
1,130 Guns.

CHAPTER VIII.

The Naval Situation.

THE fighting in East Prussia in 1914 can hardly be dismissed without a word upon the Naval Situation. Both Russia and Germany possessed large sea-going Fleets, and the operations of their Armies were carried out close to a coast-line accessible to either. One might think, therefore, that here was an opportunity for co-operation between the two Services. Russia made no effort to employ her Navy in aid of her Army, and the attempts of the Germans in this sphere can hardly be termed anything but amateur.

In regarding this question, it must not be forgotten that both Germany and Russia were essentially great Land Powers. The growth of Germany's Fleet was comparatively recent, and Russia has never been regarded as a great Naval Power. Co-operation between the Navy and Army has been one of the peculiarities of the Campaigns of the British Empire. Our history is one long series of over-sea operations, in which the Navy has supervised the transportation and disembarkation of the Army under every conceivable condition of war. None of the other Great Powers have had that experience. Combined Operations are admittedly the most difficult of all operations of war. They cannot be hastily improvised. The best of ships, the best of troops, previous experience and practice, combined with the closest co-operation between the Commands and Staffs of both Services, are necessary to give a reasonable chance of success.

Germany's unreadiness for such combined operations will be a surprise to those who thought that she was well prepared to undertake a descent upon our coasts.

As far as Germany was concerned, the reasons for this lack of co-operation on the part of the Navy were fairly potent. The Baltic is divided from the North Sea by the Kiel Canal, the passage of which by a single ship took 14 hours. The passing of any large Naval Force from one sea to another was therefore a lengthy proceeding. In 1914, the larger German ships had to be lightened in draught to 28 feet in order to allow of their passing through the Canal, which further delayed matters. Opposed to the whole strength of the British Fleet in the North Sea, the German Naval authorities could not risk even a temporary diminution of their Forces in the main theatre of action, by the detachment of ships to the Baltic. Unless she did so, they must remain inferior to the Russians, who had no other objective than operations in the Baltic.

The German Base at Kiel was an excellent one, but lay 770 miles from Kronstadt, and the remaining ports—Swinemünde, Stettin, Danzig, and Pillau—were suitable only for light forces.

Opposed to a superior Russian Fleet, Germany was compelled to act on the defensive on the sea.

Added to this, she was acting defensively on land in the early stages of the East Prussian Campaign. When she later transferred troops from West to East, the efficient railways were a surer and quicker method of transferring them than by sea, owing always to the menace of the superior Russian Fleet. From the Battle of Tannenberg onwards, Germany did make some efforts at Combined Operations, but her hopeless unpreparedness through lack of general direction, insufficiency of material and experience, brought these efforts to nothing.

As regards Russia, the situation was different. Russian Military Operations contemplated an advance as far as the Vistula, with the prospect of land operations against both Königsberg and Danzig. Her Fleet was greatly superior to anything the Germans were likely to leave in the Baltic, although there was indeed the temporary danger of a transfer of the German Forces from the North Sea. But offensive Russian Naval Operations in the Baltic would have complicated Germany's Naval Situation *vis-à-vis* Great Britain, and would have benefited the Allied Operations as a whole. As we know, the original Russian Military Operations against Germany were to have been defensive, and the Russian Fleet evidently received the same orders. When the Russian plan was changed to offensive action against East Prussia, the Russian Naval authorities were unable to adapt their programme to the new conditions. So little did the Russians depend upon their Fleet, that they looked upon it as an immediate defence-force for the capital, Petrograd. Admiral von Essen, the Naval Commander-in-Chief, was even subordinated to the Commander of the Sixth Russian Army, and Naval Orders emanated from the Russian Generalissimo, the Grand Duke Nicholas. The Grand Duke had apparently no idea of using the Navy to co-operate with Jilinsky's Advance, and any offensive Naval action suggested was rigorously discouraged and indeed forbidden.

The Russian Naval Base lay inside the Gulf of Finland, the entrance to which from Hannö to Moon Sound is only 30 miles broad. Inside lay the very strong position of Kronstadt, 210 miles back from the mouth of the Gulf. Near the mouth lay the Naval ports of Revel and Helsingfors, facing each other on either side of the mouth. Both Libau and Riga had been abandoned after the Russo-Japanese War as being too near to the German Frontier for security.

In the Summer period with which we are dealing ice-conditions did not enter into the problem.

Speaking generally, the situation of the Bases of both Germany and Russia favoured a Naval Defensive.

At the outbreak of hostilities, the Germans, realising their limitations, ordered their Naval Forces in the Baltic to conform to the following Instructions :—

(i) The principal object was to interrupt as far as possible any offensive operations which might be undertaken by the Russians. Kiel Bay was to be secured against British and Russian Forces. Enemy commerce in the Baltic was to be damaged.

(ii) Mining operations against the Russian Coast were to be instituted as soon after the outbreak of war as possible.

(iii) Portions of the High Seas Fleet, to deliver a blow at the Russian Fleet, would be detached to the Baltic according to circumstances.

Within a few hours of the outbreak of hostilities, the two small cruisers "Magdeburg" and "Augsburg" appeared off Libau, which they shelled, as well as lighthouses along the coast. On 2nd August, it was found that Libau had been abandoned even by light craft, the Russians going so far as to blow up the ammunition dumps in the town.

The entry of Great Britain into the war was electrifying. The Germans withdrew all their Baltic forces close in to Kiel, and took every precaution to render the harbour safe against British attack through the channels between Sweden and Denmark.

The Russians took no advantage of this situation to gain permanent control of the Eastern Baltic, making no attempt to emerge from the Gulf of Finland.

This inactivity of the Russians raised the spirits of the Germans once more, and they recommenced their mining operations in the Eastern Baltic. Many of their raids were pushed right up into the Gulf of Finland, but nothing could tempt the Russians out of their base.

On the 17th August, the date upon which Rennenkampf crossed the Frontier of East Prussia, the Russian Fleet emerged in force. They found the two small cruisers, with three torpedo-boats, and the "Deutschland," a mine-layer, cruising off the mouth of the Gulf. The discovery of the mines laid by the Germans was sufficient to deter the Russians, during the 17th and 18th, from carrying out any offensive operations.

The Germans continued their raiding and mine-laying operations right up to the 25th, the eve of Tannenberg, without opposition. On the 25th, the "Magdeburg" grounded close to Odensholm, well inside the Gulf of Finland, and had to be abandoned.

It is not suggested that the Russians, with no experience of combined operations, should have attempted a landing on the German Coast, but they certainly could have shelled Pillau and Danzig and have destroyed many lighthouses, besides interrupting the German trade with Sweden. Russian offensive Naval operations would certainly have had a great moral effect upon the Germans, and the passivity of the Russian Fleet under the circumstances was inexcusable.

With the coming stroke at Tannenberg, which it will be remembered commenced on the 26th August, and to counteract the loss of the "Magdeburg," the Germans now passed the IVth Battleship Squadron of 7 ships into the Baltic, but since the Russians still made no effort, the Squadron soon returned to Kiel. The success at Tannenberg relieved the German situation considerably.

With the successful operations against Rennenkampf in September, the Germans conceived the idea of Naval co-operation.

The following order was issued by the German Naval authorities :—

"The First Russian Army is retreating Northwards *viâ* Memel. It is of extreme importance to hamper their retreat from Memel and the Kurisches Haff, and to assist our Army by destroying the Memel River bridges up to and beyond Tilsit. Assistance in transporting Danzig troops in any available craft and by aircraft from Putzig is also to be rendered."

Admiral Behring, commanding the German detached Naval Force in the Baltic, took some hours in getting into touch with Hindenburg's Headquarters. Admiral Behring was of opinion that Naval craft could not reach the bridges without being sunk, and the light cruiser "Amazone" and a torpedo-boat were despatched to Memel. Hindenburg had received no orders for embarking troops, and by the time the conversation between him and the Admiral had been established, it was discovered that Memel was out of danger. Hindenburg suggested that the German vessels should destroy the Russian vessels in the Niemen, but this was judged impracticable. The landing-party of the "Amazone" was put ashore at 6 P.M. on the 14th September at Memel, but was re-embarked the next day.

On the 19th September, General von Falkenhayn, the Chief of the German Staff, requested the Chief of the Naval Staff to order a Naval demonstration, accompanied by a feint landing of a Brigade off the Courland coast, between Libau and Windau.

Strong forces from the High Seas Fleet were placed

under Prince Henry's command for this operation—namely, the 4th and 5th Squadrons of old battleships, the armoured cruiser "Blücher," two new light cruisers, two destroyer flotillas, and a mine-sweeping division; whilst the 3rd Squadron, consisting of the newest battleships, was to be employed when the two armoured cruisers, which had been detached from the High Seas Fleet for operations in the Baltic, had rejoined the fleet. The Battle of the Marne had just been fought, and the Battle of the Aisne was raging, and in place of weakening the High Seas Fleet almost to the point of disappearance, energetic action by the German Navy in the North Sea would appear to have been more than ever desirable at the moment, rather than such a demonstration in the Baltic.

The Naval Operation Order was as follows :—

"Proceed to Windau with the forces mentioned and with any Baltic forces that are available. Make preparation for a conspicuous landing of a large number of troops. The brigade is to be embarked at Neufahrwasser (Danzig) as soon as possible and to follow under convoy. Transports must be collected from the Baltic, and, if necessary, from Hamburg."

Feigned operations only, on a large scale, were contemplated. Six battalions of infantry, 3 batteries and 2 squadrons, with transport, were to be embarked. As, however, the transports could not be ready in less than a fortnight, the proposed landing was abandoned, and the operation was confined to a demonstration with the naval forces available. Endeavours were made by means of conspicuous movements, such as mine-sweeping and reconnoitring, to give the impression of an intention to land a large body of troops at Windau—an impression which it was hoped to strengthen by continuing the fitting out of the transports.

The operations were interrupted by an alarm on the night 24th/25th September that the British Fleet had penetrated into the Great Belt after destroying the mine-fields by old ships, whilst almost simultaneously a British submarine was reported in the Kattegat. The former report was falsified during the afternoon of 25th September, but the German Naval Commander-in-Chief adhered to his decision to break off the operations and return to the Western Baltic, because he had become convinced that Windau Harbour was unsuitable for the landing of troops, and the ships could not give effective support.

The slight impression made on the Russians by the demonstration at Windau was out of all proportion to the demands made on the German personnel and matériel and to the risks taken. The Russian Official telegram of 29th September said :—

"At the commencement of the war enemy ships were sighted repeatedly off Windau. At 4 P.M. on 23rd September, 18 destroyers and 1 cruiser approached fairly close to the coast and a German destroyer was sighted simultaneously off Backofen Lighthouse, about 13 miles to the westward of Windau. The destroyer lowered boats to take soundings, but these were driven off with great loss by Russian gunfire. The destroyer then bombarded the lighthouse and withdrew. On 24th September, 38 German vessels were sighted off Windau; they included 9 large warships and 7 transports. They confined themselves to a little manœuvring and then withdrew in a southerly direction. The destroyer captured a fishing vessel and destroyed her after transferring her crew."

This report gives no indication that the Russians feared a landing. It must be concluded that the repression of Admiral von Essen by the military authorities was responsible for the immunity enjoyed by the German force during the operation.

During the operation only the western entrance of the Gulf of Riga and the area west and east of Gothland had been guarded by weak forces, whilst the Gulf of Finland, the main gateway of the Russian Fleet, was left unwatched during the whole operation.

On the 25th September, the day the operation was abandoned, some 700 troops had been embarked by ships of the 5th Squadron, with a view to landing at Windau. A report by the Captain of one of the battleships of the 5th Squadron says :—

"The landing-party sent by the Army consisted of 700 young Reservists who had had no infantry training or rifle practice, and were not properly equipped. They had no cooking utensils, water-bottles, kit-bags, or tents. The absence of ordnance survey maps was a further drawback to their officers. At first only a feint landing was intended, but afterwards instructions were given to destroy the railway from Windau as far inland as possible; the small landing-party would therefore have been obliged to break up, part of it remaining to cover the retreat. In addition, it was unlikely that a town of 20,000 inhabitants would have been left without troops, so that even if the German landing-party had been able to disembark under cover of the ships, they presumably would have met with resistance when penetrating farther inland or entering Windau, and their withdrawal might have been accompanied by some losses. The whole of the 5th Squadron's landing-party would probably have been obliged to disembark to cover the retreat."

CHAPTER IX.

Retrospect and Lessons of the Campaign.

1. The Effects of the Campaign.

The Invasion of East Prussia ended disastrously for the Russians. The Allies and Russia herself had entertained great hopes of success. In the short space of fourteen days these hopes were rudely destroyed by the result of the Battle of Tannenberg. Much has been written to prove that this early action on the part of Russia, carried out to the detriment of her own interests, saved the Allies in the West, until it has gradually been assumed that this is the case. The Russian Generalissimo, upon receipt of the news of the disaster to the Second Army, turned to the French Military Representative with the words, "Nous sommes heureux de faire de tels sacrifices pour nos Alliés." He thereby implied that the Invasion had served its purpose. What are the actual facts supporting this contention ?

It has already been shown how the Russians had been unable to keep up to their engagements, entered into in time of peace, to carry out an immediate offensive against the Germans, but how, under pressure, they had tried to improvise a fresh campaign in order to approximate to these engagements. It must be accepted as an undoubted fact that the Germans were well-informed as to the probable numbers and dates of concentration of the Russian Armies, and as a consequence of their calculations they had decided to reduce their forces in the East to a minimum, in order to secure the necessary superiority for a decision in the West. However cautious a policy they may have attributed to the Russians, they can have been under no illusions as regards the fact that the Russian Army would move against East Prussia as soon as it conveniently could, and that the weak German forces left in the East could put up no permanent defence against the Russian Forces which might be brought against them. In their general plan, therefore, they had earmarked six Corps, two each from the right, centre, and left of the line in the West, for employment in the East should the situation demand it. These six Corps were not considered as the additional force necessary to achieve a decision in the East, but as the maximum force which might be necessary to maintain a reasonable equilibrium there, and to be drawn upon as required.

The early Russian advance from the Narew came as a complete surprise, and under the strain of the news the German Commander in the East collapsed, and had to be replaced.

It was on the 22nd August that Ludendorff was appointed Chief of Staff to the German Army in East Prussia, and we have it on his authority that the German Supreme Command made no mention of large reinforcements for the East at this date.

Ludendorff did not ask for any. The only units set in motion for the East were the 1st Landwehr Division from Schleswig-Holstein and detachments from the Vistula fortresses. The movement of these troops had no influence upon the Campaign in the West. So far, then, the early Russian advance had created a temporary panic in East Prussia, considerable outcry in political circles in Berlin, and annoyance at Supreme Headquarters.

The situation in the West on the 21st August, when the news of the collapse in East Prussia arrived, was as follows: The French offensive North and South of Metz had definitely been brought to a standstill, and the main German advance through Belgium was proceeding according to plan. The First German Army had passed Brussels and Hal, shouldering the Belgians aside to Antwerp, but no contact had been gained with either French or British. The Second German Army had reached a line from just South of Waterloo to the Sambre South of Gembloux. The bombardment of Namur had commenced, and the heavy howitzers already tested at Liège promised a quick reduction of the fortress. No contact had been gained with the French. The Third German Army had reached the Meuse from Namur to Givet, having gained contact with French Infantry and Cavalry opposite Dinant.

Thus, three German Armies, totalling 760,000 men, were well on their way through Belgium to France. They were apparently only opposed by the Fifth French Army, totalling 254,000 men, and the despised British Army, if indeed that had arrived on the scene. All had gone well with the plan in the West, but it was early to say what the outcome of the great movement was going to be. The collapse in the East was not considered so calamitous as to necessitate the detailing of troops from the West in order to save the situation against the Russians.

By the 24th August, the majority of the forts of Namur having fallen, the XIth and Guard Reserve Corps, both of which had formed part of Von Gallwitz's force for the siege of the fortress, were released for other duties. At 2 A.M. on the 25th August these two Corps were ordered to the inner flanks of the Second and Third German Armies respectively. It was not till about mid-day on the 25th that these orders were altered and the two Corps ordered to march off for entrainment for the East.

It may therefore be assumed that the decision to reinforce the East was taken between 2 A.M. and 12 noon on the 25th August.

News arriving from the East during the period from the 21st to the 25th had been hopeful. Rennenkampf had done nothing to follow up the retreating Germans, and the new Commander in the East was about to commence an offensive against the Second Russian Army. He had made no demand on Supreme Headquarters for reinforcements. Indeed, we have the testimony of General Hoffmann, the officer directing the Operations section under Ludendorff, that the Chief of Staff expressly stated on the telephone to General Tappen, the officer directing the Operations section at Supreme Headquarters, that he did not require any reinforcements, when informed that it had been decided to transfer troops to the East.

What was it, therefore, which decided Supreme Headquarters to carry out this move? There was certainly no lack of organisation at Supreme Headquarters, where it must have been known that no troops being despatched at that date could take part in the operations just about to commence against the Second Russian Army.

General Tappen gives the reason as follows :—

"In this period, along the whole line of the German Army on the Franco-Belgian frontier, a more or less closely connected great battle had taken place in a short space of time. The French, as had been expected, had offered battle, to prevent us from penetrating into France, and they had been beaten in this great continuous battle. The exceedingly favourable news which came in daily, and also on the 25th August, in combination with the great victories achieved by the Sixth and Seventh German Armies in Lorraine from the 20th to the 23rd August, aroused the feeling at Supreme Headquarters that the great decisive battle in the West had already been decided in our favour. Under the impression of this decisive victory, the Chief of the General Staff, in spite of the considerations against doing so, came to the decision to send troops to the East. He believed that the moment had come, in continuation of the general plan of operations, to send troops to the East in order to seek a decision there also."

General Von Molkte, in a memoir written during the summer of 1915, gives a different reason for his action. He says :—

" While the first five German Armies were engaged in advancing victoriously over the Meuse and the Sambre, the conditions in the East, where the Russians had advanced unexpectedly early, made it necessary to send reinforcements to that theatre before a decision could be reached in the West. I intended to take these reinforcements from the Seventh Army, which, as was the case with the Sixth Army, despite heavy fighting, had been unable to advance far on the Moselle. The definite reports that the enemy opposing both Armies remained in superior numbers, and that their own losses had been so great that further employment of the Seventh Army was only possible after units had been brought up to strength, was the cause, after the fall of Namur, of my taking two Corps from the German right wing and of my sending them to the East. I admit that this was a mistake, which had its revenge upon us at the Marne."

At the moment the order was issued, such was the confusion upon the German Lines of Communication that the three Corps detailed had considerable distances to cover towards the rear to reach their entraining areas—Malmédy–St Vith, Aachen, and Diedenhofen respectively. It was thus not till the morning of the 27th that the XIth and Guard Reserve Corps commenced entraining.

By the morning of the 27th the situation on both fronts had improved greatly. In the East, the Battle of Tannenberg had commenced without interference on the part of Rennenkampf. In the West, the Battles of Mons and Le Cateau had shown that the French and British could not hold up the advance of the German right wing.

Von Molkte had thus had from the evening of the 20th till the morning of the 27th in which to consider his decision, which cannot therefore be classed as a panic decision. Immediately he realised the extent of victory of Tannenberg, he cancelled the entrainment orders of the Vth Corps, which was still waiting for rolling-stock at Diedenhofen. It is probable, therefore, that the reason given by Von Molkte is the correct one, for, had he been of opinion that the decision had already been gained in the West, he would have been sending more troops to the East and not countermanding the move of those already ordered there.

Thus the final effect of the early Russian Advance was the withdrawal of two Corps from the West. Can it be argued that this saved the Allies from disaster? The three Corps were withdrawn on the 25th, and the Battle of the Marne did not take place until the 9th September, fifteen days later. Von Molkte had ample time in which to replace these Corps in his right and right centre by troops from his left, or to make a general move to his right, had he so wished. That he did not do so points to the fact that he was quite satisfied with the progress being made by his right. Moreover, the presence of two more Corps in his right wing would not have placed it in a position to include the fortified area of Paris in its swing, and it was the existence of this area, where the French forces could manœuvre but where movement was denied to the Germans, which made it possible for Galliéni to outflank the

German right North of Paris. A far larger reinforcement than two Corps would have been necessary for the inclusion of Paris in the German advance.

Had the Russians waited until their forces were really ready—that is to say, until the 20th day of Mobilisation—they would have advanced into East Prussia some six or seven days later than they actually did. There can be no question that they would have been able to maintain a more continuous pressure from the commencement of their movement, and that their men would have fought better at Tannenberg. As events proved strategically, their advance must have had the same strategic effect, and might have been even greater when we consider what the unpreparedness meant to the fighting value of the rank and file.

Russian casualties during their twenty-eight days' Invasion of East Prussia amounted to 310,000 men and 650 guns, a force equivalent to 7½ Corps. When it is remembered that these casualties were inflicted on the flower of the Russian Regular Army, the extent of Russia's loss can be imagined. These casualties were inflicted by a German Army, which at no time exceeded 7½ Corps in strength, upon a Russian Army of 12½ Corps with 8½ Cavalry Divisions, and within the space of twenty days, the very speed of the operations enhancing their moral effect. Beyond the loss of prestige inevitable in such a defeat, there was lost also a vast amount of material which could not be replaced at any time during the War. The losses in officers and non-commissioned officers were never made good. The subsequent history of the Russian Army, up to its final eclipse in the Revolution, was one long story of attempts to make up for losses incurred in the first few days of the War.

As regards the moral effect of the Campaign, the Germans also greatly benefited. The victory of Tannenberg, followed closely by that of the Masurian Lakes and the freeing of East Prussia, eclipsed the German defeat on the Marne, and enabled the German authorities to conceal the real meaning of the Marne—the failure of a short and glorious campaign—amidst the general enthusiasm over the defeat of the Russians.

One must come to the conclusion that the evils wrought by the unprepared advance far outweighed the little good it may have temporarily caused.

2. Military Plans.

A Military Plan must be elastic. The conditions under which the Military Forces of a Nation may have to be employed are continually altering. Statesmen cannot foretell the exact conditions under which a war will be waged, and yet the Strategist must envisage all eventualities, and be prepared to adapt his plan to the situation of the moment. Nations group themselves in armed alliance, and co-operation between nations will be the rule rather than the exception. Plans elaborated in peace must be prepared in great secrecy, and this secrecy will militate against successful execution, since so many important individuals and interests cannot be consulted, and any of these may come forward at the last moment and insist upon important modifications.

Statesmen are not strategists, and busy men as they are, they are apt to pass over some point when agreeing to some plan in peace time, upon the alteration of which they insist when the crisis is upon them. The Campaign envisaged by the General Staffs of France and Russia was based upon a simultaneous advance against the main enemy, Germany. The Russian Army had failed to make the necessary preparations in peace time, and the French found themselves opposed by a larger force than had been expected. The French offensive was checked in its early stages, and the initiative on the Western Front was lost for the moment. The French plan was not elastic enough to meet the great threat coming through Belgium, with the result that much precious territory was lost for four years.

Military plans must also be psychologically sound. In the seclusion of a quiet room it is possible to forget the question of psychology altogether. A free people will not willingly submit a portion of their territory to foreign invasion, even to ensure success in the war as a whole. Popular outcry will insist upon the alteration of a plan embodying such an eventuality, with consequent danger to the general situation.

The German General Staff had decided deliberately to risk an invasion in East Prussia, in order to secure a decision in the West. Probably the people of East Prussia had no inkling that they were being sacrificed for the general good, and they at once began to show restlessness at the prospect of a barbarian inroad into their country. The calculations of the German General Staff led them to believe that the decision in the West would be a quick one, and that they would be able to return and face the Russians before the damage was irreparable. This did not prove to be the case. Streams of refugees from the favoured province entered Berlin, and there was an outcry against the plan. There arose a body of feeling in the capital which insisted upon more energetic action in the East, and fresh troops, as they were collected, were hurried against the Russians, any superiority in the West being finally prevented. Military thought cannot afford to ignore the ordinary feelings of human nature, and the German plan of ignoring the Russians in the initial stages of the war was false in principle. The great fault of the German General Staff throughout the war was that they placed themselves apart from the German people and ignored their feelings.

Military plans must be administratively sound. There is a tendency to despise administration in all Armies. Administrative duties are undoubtedly wearisome to some brains. They involve the consideration of much detail, and they do not deal so much with the human element. Administrative Staff Officers do not come so much in evidence with the men, their duties being often regarded as non-fighting duties, and to be despised in consequence. In modern war no greater mistake than this can be made. The Army with which a Commander carries out his plans is a composite weapon. It requires not only to be fashioned correctly for its purpose, but to be used within its capabilities. The only Commander of the future who is likely to succeed is he who suits his plan to his weapon, and to do this he must evidently know the physical as well as the psychological limits of that weapon. This has always been true, but in the main this truth has been overlooked in the past, because the size of the Armies was well within the physical capacity of the theatre of war. Let an Army increase, however, to a size which puts a strain upon this capacity, whether it be of movement or of maintenance, and at once the limitations in the use of the weapon become apparent. When Armies such as those of Germany, Russia, France, and Great Britain are concentrated in areas restricted in size such as France, or in development such as Russia, these physical limitations assume paramount importance. The success or failure of a Commander of the future will depend upon the degree to which he recognises this fact, and makes allowance for it in his plan. It is not enough that he may be supplied with expert advice. He must himself be capable of appreciating the relative value of all the factors which affect his plan, and in nearly all cases these factors are those of administration.

The principles of war are few, but, as the text-book tells us, their application is difficult. It is easy to read up in a book how two Armies may converge upon one another, and how various advantages and disadvantages may accrue from interior and exterior lines. But what is the good of issuing orders for such manœuvres unless they are feasible? Can the distances be covered within the space of time allotted, and will the roads and railways bear the increased weight of traffic demanded? When the troops have reached the desired spots, can they be maintained there in ammunition, food, and clothing ?

The Russians had neglected Administrative training in their Army. The attitude to his Administrative Staff of a Russian Commander was, " Here is a plan that I have made. It is your business to see that it is carried out." It must be acknowledged that the Russian conception of a quick advance to help their Allies was a fine one, yet it meant advancing without the necessities of war. It meant misery and suffering for the men in their struggle to keep up with an impossible time-table of marches, and it meant their final defeat in battle through lack of means to maintain their strength.

The Germans had perfected their Administration to a high degree. Their Lines of Communication in France and Belgium had been carefully worked out in minute detail, and yet their strategy ran away from their ways and means. Their men were over-marched, and they were ill-supplied with food. In East Prussia there were many examples of failure in the German administration. The night-march of the XVIIth Corps on the night of the 19th/20th August, to reach the field of Gumbinnen, was badly arranged. There was no proper traffic control to clear the roads of refugees, and all calculations of time and space were upset. The crossing of the Lötzen Gap on the 8th September by the XVIIth Corps and two Cavalry Divisions was another administrative failure.

Improvisation on a large scale in modern war is becoming more and more difficult owing to the administrative difficulties involved. The British

Campaign in the Dardanelles is an example of the truth of this assertion. Such improvisations are better left unattempted if there is any doubt about the soundness of the administration.

3. Envelopment.

Graf von Schlieffen, the Chief of the General Staff in Germany from 1891 to 1906, had made a deep study of war from history, and in the later years of his appointment from the reports of his observers in the field during the Russo-Japanese War. His views upon modern war permeated the German General Staff from top to bottom, and all the training of the German Army was based upon these views.

Germany's position in Europe made it essential that any campaign in which she took part should be short and decisive. Von Schlieffen considered that though frontal attacks might still succeed under modern conditions, yet the results to be gained by them must always remain small. He had therefore devoted his attention to the study of campaigns in which envelopment of the enemy had been achieved, and had modelled the plan for all future German campaigns upon them.

He was of opinion that the day had gone by when envelopment could be commenced *after* contact with the enemy had been gained. It was not therefore economical to maintain reserves behind the front line of an advance, but was preferable to deploy every available man from the outset, so that he could take part in the first great engagement, which thereby became decisive. The conception of envelopment must be aimed at from the very outset of a manœuvre. The increasing size of modern armies necessitated their deployment over a wide area in order to make use of every available road and railway. It therefore became necessary to give these widely extended forces some geographical point or area towards which they were to converge. We see this idea in the direction given by Von Molkte to his two Armies for their advance into Bohemia in 1866. This area was by no means rigidly fixed, for the ideal point of convergence naturally depended upon the movements of the enemy, and might alter from day to day as the mutual situation of the advancing wings changed. Von Schlieffen believed that if he could achieve strategic surprise, he would be able to herd the enemy into the area he had chosen beforehand. The conception was, in fact, the advance of a non-rigid line, aimed towards the enemy, with one or both wings ready to turn inwards, when the enemy had been located by some portion of the centre of the advancing line.

Von Schlieffen realised that the conception of envelopment based upon such a method was only suitable to the army with a superiority in numbers or mobility, but he calculated that by choosing the proper moment he could steal a march on his enemies which they would be unable to catch up.

The configuration of the frontier, limiting the initial deployment of the forces detailed for the advance, naturally affected this idea of envelopment from the commencement of the manœuvre.

The Russians were exceedingly well situated as regards East Prussia, whereas the Germans were not so well placed as regards France. To avoid the fortified frontier of France they had to send their enveloping wing through Belgium, entailing a long and tedious march to reach a position from which they hoped to be able to outflank the French left and drive the French forces up against their Eastern frontier. In Ludendorff's two operations the strategical situation differed in the same way. For the Tannenberg operation his troops were initially well placed, could he but disengage those opposed to Rennenkampf, whereas for the Masurian Lakes his right wing had to carry out a long march to reach the required strategic position.

As regards numbers, in their East Prussian Campaign the Russians had a superiority of 5 to 3 in Infantry and 3 to 1 in Cavalry. In the advance in the West, the Germans had a superiority of 4·2 to 3 in Infantry and a practical equality in Cavalry. In Ludendorff's two operations, the Germans were much nearer equality in numbers, but greatly superior in mobility to the Russians.

Out of these four operations, that of Tannenberg was the only one where envelopment was achieved.

The Russian initial advance into East Prussia failed owing to bad leadership and bad administration. They failed to bring their two armies into tactical supporting distance of one another, and they brought their men in an exhausted state to the battlefield. As regards the Germans in the West, their enveloping movement failed through lack of numbers. They were unable to include Paris in their swing, and were outflanked themselves instead of outflanking the enemy.

In Ludendorff's two operations there were critical moments when successful envelopment hung in the balance. At Tannenberg, on the 28th August, the XXth Corps in the centre being in difficulties, Ludendorff ordered his right wing under Von François to close Northwards to Lahna to prevent the defeat of his centre. Von François deliberately disobeyed this order, and maintained his initial direction of Neidenburg. Had he obeyed Ludendorff's order, the Second Russian Army would have escaped. Von François had achieved complete envelopment, and his weak detachments along the Neidenburg–Willenberg road and railway were able to maintain their positions; but had the the Russians retained any control, these detachments might have been in a difficult situation. As it was, the Russians had been exhausted by their efforts in the advance, and could put up no sort of offensive. Again, during the Masurian Lakes operation, Ludendorff ordered Von François to close in Northwards to the support of the XIth Corps in the centre. This time the order was obeyed, and the First Russian Army escaped from the jaws of the pincers.

Every operation cannot be a Sedan. As Clausewitz says, "It required a Varro to make a Cannæ." Later, it required a Jilinsky, a Samsonoff, and a Rennenkampf to make a Tannenberg.

It may be that the advent of mechanical vehicles will make envelopment easier in the future, but the fact remains that considerable superiority is necessary to achieve complete envelopment of a large and mobile enemy army. The greatest danger to which a line advancing in this way with a view to envelopment is exposed, is defeat in detail. The proportion of troops to be allotted to the centre and to each wing must usually be fixed from the outset, and there will be little opportunity of making any material alteration once the advance has commenced. In long advances such as those through Belgium or against the Masurian Lakes, an enemy who keeps his troops well in hand will have time to make up his mind as to the strength of the various portions of the line advancing against him. If the wings of the advance are weak they will not achieve envelopment, whereas if they are strong and the centre weak, the enemy, from his central position, may break the centre, thus separating the two wings and defeating them later in detail.

As regards envelopment in minor operations forming the detail of a battle, we find numerous examples in the German operations. Against Sirelius at Neidenburg on the 30th August, at Arys, at Lyck, and at Soltmahnen, during Von François's movement to the East of the Masurian Lakes, the German tactics of envelopment were completely successful.

4. Nationality.

The sense of Nationality is one of the strongest factors in modern civilisation. Oppression appears but to increase this sense. Both Russia and Austria-Hungary had failed to absorb their subject Nationalities, though they had resorted to every means of oppression, mild or violent, in their efforts to create a homogeneous whole. The British Empire has achieved better results by the granting of freedom within the State to all Nationalities forming part of the Empire. The most striking example of the success of this policy may be seen in the Union of South Africa. The Boer War of 1899–1902 finished only twelve years before the commencement of the Great War, and the Union was declared four short years before it. Had this freedom not been granted when it was, the risings in South Africa would have been much more serious, and might even have prevented us from developing our full Imperial strength on the side of the Entente. As it was, we found loyal subjects who succeeded not only in restoring law and order, but also in conquering the German African Colonies, and in sending a considerable contingent to help us in France. Such a system of political freedom undoubtedly produces the best fighting material, and it is on the fighting material that the eventual result of a great struggle rests. The Russian units were diluted with one-third subject races, just as were the Hungarian units, and their fighting capacity was far lower than that of the German units.

The question of Nationality is an important one for every State. The Dominions of the British Empire fully realise this, and are demanding immigrants of British stock in preference to those of foreign extraction. Many of the newly constituted States of Europe have not realised this cardinal factor for their future safety. They are ever ready to absorb fresh territory and with it alien races.

If, after the first feeling of hostility has subsided, they do not grant these aliens a large measure of political freedom, they will find that the subject races they have created are a weakness in peace and a danger in war.

Every Nationality has its own characteristics. It is a century since Britishers were ranged alongside foreigners in war, and many of our Officers had forgotten that to get the best results out of such Allies it is necessary to study their characteristics. There were many cases in the late war where incompatibility of temperament failed to get good results, and happily many brilliant examples of sympathy and understanding. It behoves our Officers to combat the national feeling of insularity, and to study the characteristics and points of view of our Continental neighbours, so that we may be able to give a good account of ourselves when called upon to act either with or against them in any future struggle. Failure to achieve results through personal prejudice is inexcusable.

5. Surprise.

The Soldier is often accused of allowing himself to be surprised in each successive war in which he engages. He is told that he has copied the methods of the last war, without taking sufficiently into account the advance of science. It must not be forgotten, however, that the Soldier has many difficulties in peace time. His trade is one which is difficult to practise without the whistle of the actual bullet. The effect of new weapons is largely a matter of opinion, since they cannot be practised against the human body. Democratic Governments, moreover, are loth to expend large sums of money on requirements for war, and consequently much that is considered necessary by the Soldier is ruthlessly rejected on financial grounds.

Our Field Service Regulations tell us that Surprise is the most powerful and effective weapon in war. Surprise is not by any means confined to the battlefield. Nations are for ever trying to prepare a surprise for their future opponents. The Germans surprised their enemies in Organisation by employing their Reserve Formations in almost the same way as their First-Line Troops, and thereby upset the calculations of the French and Russians by the increased numbers appearing so early in the field. Napoleon surprised his opponents by decentralising his command and giving increased responsibility to his subordinate Commanders, thereby increasing the mobility of his columns. Future surprises may be obtained through a combination of road and railway transport, perhaps even of air transport. The surprise may be again one of material. The French produced the mitrailleuse, and later the '75 gun. The Boers surprised us by producing heavy guns in the field. In East Prussia the Germans surprised the Russians by their Heavy Artillery fire. The mobility also of the Germans produced the impression of ubiquity amongst the Russian soldiery, who considered themselves inferior in numbers in consequence.

Nations watch each other carefully in peace time. Their Intelligence Departments are ever awake to the production of a new weapon. The invention of new weapons, however, is seldom the production of a Military Department, and such inventions are therefore difficult to keep secret. If new weapons are not thoroughly tested, they may be but a snare and a delusion, for one can imagine the *moral* of the men falling at the discovery that a much-vaunted new weapon has not the effect predicted of it. It is quite within the bounds of possibility that an unscrupulous nation may seize the moment that it has stolen a march on its neighbours, by the invention of a new ray or gas, to make an unprovoked attack before the new secret is out. Intelligence Departments must therefore never relax their efforts.

It is, however, rather in new tactical methods, built up round the employment of a new weapon, that an initial surprise may be created. Such methods may well escape the eye of even a trained military observer, and may be of decisive effect in the opening stages of a war. Our Musketry surprised the Germans at Mons and Le Cateau. The Russians had been taught to depend upon weight of numbers, and were surprised by the effect of modern weapons. As the Great War proceeded, surprise succeeded surprise, each in turn being countered by an antidote. The heavy preliminary bombardment produced the elaborate trench system and the deep dug-out. Trench systems were then overrun, and the deep dug-outs cleaned up behind. Then came the pill-box and machine-gun nest, which were blinded and neutralised by smoke-clouds. All these surprises followed in quick succession under the stimulating effect of war.

The military mind is infinitely more sluggish in peace time. The Russo-Japanese War of 1904 had shown that wire was the limiting factor in the Infantry Advance, and yet no antidote had been discovered by 1914. This important stage had undoubtedly been conveniently glossed over.

We must strive our hardest to produce a sound tactical training in the Army. It is the duty of the General Staff to see that a keen imagination is brought to bear, so that a proper picture of future war is presented to the troops they are training. This picture must be revised continually, and perfected by reading, thinking, and experience. No stage must be omitted, because the solution is difficult. Junior Officers must be encouraged to put forward their difficulties, for it is they who will make the unpleasant discoveries. Step by step these difficulties must be solved, and the whole worked up into an efficient system of war.

The great essential in a modern Army is Education. It is the educated mind, contrary to the expectation of many, which adapts itself best to the strain of war. The Russians were largely uneducated, and at a great disadvantage to their opponents, the Germans, and their capacity for fighting was lower in consequence.

6. The Course of Future Wars.

In 1914 the general belief prevailed amongst Military authorities that a modern war would be of short duration. No nation had prepared for a long struggle. All efforts had been spent upon developing a quick Mobilisation and concentration, followed by a tremendous blow calculated to sweep everything before it. The decisive battles were expected to commence about the fifteenth day of Mobilisation. A war of about the duration of the campaign of 1870 was predicted, financial authorities being generally of opinion that European Nations had become so interdependent that none could bear the strain of a long war.

Germany, France, and Russia each delivered a great blow. In each case, and in varying degrees, the force of the blow was met by the opponent and parried. Static warfare then ensued, owing to the superiority that had been gained by the defensive, and the struggle became a battle of resources in men and material.

The story of the failure of the Russian blow has been told. The German blow very nearly succeeded. Its failure was due in a lesser degree to lack of good leadership, and more to the fact that the strategical conception was too great for the administrative possibilities of the situation. Had these blows succeeded, the war would not have been a long one.

After the experience of the late War every nation has carefully taken stock of its national resources, so that it may not be caught napping once more by a long campaign. No nation will risk again going to war without preparing for a long one, but that is by no means to say that future campaigns will be long ones. In fact, with the terrible results of a long war upon both conquerors and conquered alike before them, we may be sure that a nation contemplating war will strain every nerve to ensure success in as short a time as possible.

The question for us to consider is whether a great blow, carefully prepared in the light of past failures, and supported by every modern invention that can be turned to military account, may not in future campaigns be carried through to a speedy conclusion. Let us try therefore to picture the progress of a future war of the nature of the last, where the main combatants are divided by land frontiers.

Each side will have a comparatively small Air Force, recruited on a Regular Basis. Behind these Regular Air Forces there will be Reserve Air Forces, drawn from the commercial aviation assets of the nation. The object of each side will be to place all its Air Forces on a war footing as quickly as possible, and to deal its enemy a great blow before he is ready. The objective of this blow will be to reduce the *moral* of the enemy, to the point even of taking away his desire to make war, and to destroy his preparations in men and material, so that he will not have the possibility of carrying on war. The attack will be delivered against the Government, arsenals, depots, communications, industries, and troop concentrations, by bombing them from the air. There are authorities who maintain that all that is required is a vast Air Force capable of carrying out such an attack, and that a war will be brought to a conclusion in a short space of time. There is no doubt that destruction of life and material can be carried out on a vast scale by the Air Contingents likely to be in existence in the future. But there are many things which tend to militate against such a speedy termination of hostilities. In the first place, nations will contest for Air supremacy in peace time, and there will be large Air Forces on both sides and severe Air fighting. Casualties amongst the fighting aircraft, as well as amongst the bombers, will be heavy. And in the second place, the element of the air is not like that of the sea. Air supremacy cannot be expected such as was achieved after the battle of Tchushima or by the British Fleet in the late war. Anti-air defence, by gunfire and dispersal of resources, will make it more and more difficult for bombers to obtain their results without great loss to themselves. Such casualties, in conjunction with the loss of Air resources on the ground in the general destruction taking place, are more than likely to bring about some kind of Air stabilisation. By this is meant that though Air Forces will still take the air on both sides, their strength will be so reduced that neither can achieve decisive success in destroying the *moral* and resources of the other, though each may still be strong enough to prevent the other from making good the casualties in men and material.

All Aircraft available, from whatever source, will be brought into play in this initial phase, and the blow delivered will undoubtedly achieve reduction of resources. This should tend to shorten future wars. But reduction of *moral* is another matter. In cases where there is racial hatred, it is conceivable that the indiscriminate killing of people in industrial areas, such as must occur in air-bombing, will increase that hatred, and with it the desire to go on fighting. It would seem that *moral* will only be affected by lack of food, either through deficiency or through lack of means to distribute it. To achieve such a state of affairs, lengthy bombing operations will be necessary.

Holding such views, then, of the nature of the results to be achieved by future Air action, it is unlikely that nations divided by a land frontier will depend solely or even mainly upon Air Forces in case of war. They will still continue to develop their Land Forces. Great blows, such as were carried out in the late war, will still be contemplated. Mobilisation and concentration will be as quick as ever, so that advantage may be taken of the situation created by Air action, the timing of the blows in the air and on the land being carefully co-ordinated so that the maximum effect may be obtained. Although mobilisation and concentration will be rendered more difficult under Air fire, the general reduction of resources in the initial air fighting will certainly give the blow on land a greater chance of success than was the case in the late war.

The modern Army must be one equipped with all the necessary adjuncts of war. It must have its own aircraft for reconnaissance and Air defence, and its mechanical vehicles to give it mobility. The distances such an Army will cover in future will be out of all proportion to what we have seen before. When the main operations on the ground start, it is considered that all efforts will be concentrated upon the struggle on the ground. The effect of an invasion upon the *moral* of a people is always great, with its hordes of refugees flying in front of it, upsetting communications and the distribution of food. Both sides will inevitably concentrate all their Air resources for the pushing on or repelling of such invasion. The Independent Air Forces on both sides, which have been bombing far afield, will close in their efforts to suit the strategy on the ground. The Invader's bombers will deal with enemy guns, communications, and troop concentrations, and prevent him from developing his defence against the advancing columns. His fighters will protect the advancing columns and their communications against enemy air action. The defender's bombers will strain every effort to destroy the communications and rear services of the invader in order to slow up his advance, and give time for concentration to meet it. His fighters will protect these bombers in their operations. In the case of the defender, Public Opinion would insist upon a cessation of distant bombing of enemy resources when faced by an Invasion, and upon a concentration against the immediate enemy. This is an added reason for the Air Forces of both sides to concentrate over the area of the land fighting, since they will inevitably gravitate towards one another in all operations.

When operating in close co-operation with Land Forces, Air Forces will not concentrate against the front of an Army, for their targets will not be suitable. Judging by the conditions in France during many of the great offensives, Air fire will be unable to stop troops advancing, but will considerably alter the state of affairs in rear of an Army. Far away to the rear, troops and transport will have to adopt suitable formations against Air fire, for nowhere will they be safe, even at night.

It has been pointed out that the initial Air blow will aim at the destruction of *moral* and resources. What will be the objectives of a modern invading Army ? Have they been altered by fresh conditions ? Our Field Service Regulations tell us that war can be brought to a successful conclusion only by the defeat of the enemy's armed forces and the destruction of his powers of resistance. So far, the Air Forces will have considerably lessened the enemy's power of resistance, but the armed forces still remain.

The objectives of the various great blows carried out in the Great War were varied. Germany intended to sweep round the French left flank and drive the French Army against the Franco-German Frontier, and there envelop it. They kept their eyes fixed upon the French Army, ignoring the Channel Ports and Paris. Had they defeated the French Army, these would have fallen an easy prey to them. The Russian Advance in East Prussia was undertaken to create a diversion from the Western Front. The British blow at the Dardanelles was the first great attempt to break the static warfare, by opening a route to Russia, by which her man-power could be properly equipped. In each case the defeat of the enemy forces was not achieved, and the blows failed in their object. Can it be maintained that Germany would have succeeded if she had taken other objectives than the French Army ?

The enemy's armed forces still remain the main objective, but the route by which they will be approached may be decided by two factors which have been brought into prominence in the late War. The first is one concerning Industrial Centres. Many important Industrial Centres lie close to the Frontier. They are tied to the fields of coal and iron, such as Briey and the Ruhr. They cannot be dispersed for safety. An enemy advance will attempt to include such areas in its progress, so as to deny them to the defender, and make it impossible for him to carry out a long campaign. The loss of Briey to France would have caused her collapse at an early date, had she not been assured of foreign material through the protection afforded by the British Fleet. The French occupation of the Ruhr has made it impossible for Germany to prepare for a successful war. The second has to do with Air action. Air action can only be carried out at limited distances. Each year these may be increasing in length, but great effect will not be obtained if the aerodromes from which aircraft are operating are at too great a distance from their objectives. At present, a distance of 200 miles is about the extreme effective distance. Modern Armies, with their power of quick movement, will try to drive back the enemy aerodromes to such a distance that there may be relief in their own country from bombing. The route chosen for an Invasion must therefore take account of enemy aerodromes and the communications which serve them. The Invader will try to arrange his Advance so as to incommode enemy Air action as much as possible.

The German and Russian Advances in the late War had almost reached a stage of exhaustion through lack of means when they were met and parried, but modern mechanical vehicles may be expected to facilitate future movement and prevent such exhaustion.

In any case, how are nations to delay such Invasions in future in order to gain time enough to meet them by counter-offensive ? One French writer has recommended the commencement of the framework of a defensive line along the whole North-East Front of France. He has calculated that 440,000 men would be required for its defence, and that its cost would be two milliards of francs at pre-war rates. The writer points out that France mobilised 3,800,000 men in 1914, and that 440,000 is only a small proportion of this number. He further says that the loss to France through the German Invasion was many times greater than two milliards of francs. In the impoverished state of the World, no nation would contemplate such expenditure. But, quite outside the question of expense, no such linear defence can be effective in reality. It is liable to be broken at any moment by a modern offensive, which has so gained in power through the advent of tanks.

All nations must, however, adopt some kind of defensive screen behind which they can carry out their mobilisation and concentration. Such a screen is all the more necessary for the nation that feels it is inferior in mobilisation power to a possible enemy, and the strength of the offensive and the greater mobility of modern armies will tend to make Armies concentrate farther back from their frontiers, thus making the position of the power inferior in mobilisation more difficult.

The first thing which strikes one in this question of delaying an Invasion is the power of demolitions to stop a mechanical Army. For many years to come, owing to the non-commercial nature of tracked vehicles, armies will depend upon wheeled mechanical vehicles for their mobility.

Even in 1914 the Belgians slowed up the German attack on Liège to the pace of their Infantry by hastily improvised demolitions. The Germans in their retreat of 1918 carried out systematic demolitions of bridges and cross-roads to such a degree that mechanical wheeled vehicles were out of the question. The systematic preparation of demolitions by a nation realising its inferior power of mobilisation appears to be the best solution to the problem of delaying the enemy, but there are many difficulties and objections to it. The geographical location of such demolitions is simple after a general plan has been made. There can be no question of a continuous line of demolitions all along the frontier, for it would then be equally impassable to the counter-offensive, and it must always be the object of a nation to make war in the enemy's country in preference to in its own. Demolitions, therefore, must be confined to certain areas, leaving others free for manœuvre, where other means for guarding the frontier must be found, such as Frontier Troops upon a more or less permanent war-footing or gas. Even with demolitions confined to certain areas, how will the Frontier population regard the matter? Any outward and visible signs of the abandonment of an area to the enemy always raises an outcry, and the prospect of deliberate destruction by their own authorities would raise the same. There are also practical difficulties. Large quantities of explosives would be required. If maintained in a "ready" position, deterioration would be high, and the yearly turn-over to keep things ready for action would be great and consequently expensive. It might be possible to arrange for storage near the points at which the explosives are to be employed, ensuring the placing of them in position by sowing ex-N.C.O.'s in small Government positions in the neighbourhood, as a permanent control in peace time. In this case there would be a danger of individual failure through carelessness or traitorous conduct.

The next thing which strikes one is the defensive power of gas, especially of persistent gases such as Mustard Gas. The employment of noxious gases has been limited by International Agreement; but apart from the fact that all nations are not parties to the Agreement, it cannot be assumed that nations will not adopt every conceivable measure to maintain their national existence.

Weighing all these considerations, the chance of success for a great blow of the nature described appears to be, however, greater than before, and there can be no doubt that nations going to war will attempt to carry them out.

The next thing to consider is whether such initial blows, having broken the enemy's armed forces, will really bring hostilities to an end. Guerilla warfare is abhorred by all soldiers, and we have numerous examples of lengthy guerilla campaigns. As a rule, they depend for their success upon the inhospitality of the region in which they are waged. Spain in 1800 was inhospitable, and the Spaniards were helped by the British Army. South Africa in 1900 was very short of communications, if not actually inhospitable. Both countries had a small population. Countries such as Germany and France could not long sustain guerilla warfare, though Russia possibly might. Still, the advent of mechanical vehicles, armoured and very mobile, and also of aircraft, has facilitated the overrunning of a country and simplified the suppression of guerilla warfare except in abnormal terrain, such as mountains, where neither can work with full effect.

It is considered that any civilised country is less likely to continue a campaign after its armed forces had been scattered than ever before.

The argument may be advanced that Air action will so reduce the resources of both sides that Land Armies will not be equipped with all their necessary adjuncts, that offensive weapons will be so scarce that the defence will again have the upper hand, and that mobility will disappear. If the will to fight still exists under such conditions, then wars may indeed become longer than ever. Countries will be devastated, and we shall come back to the conditions of the Hundred Years' War. Is such a picture possible, with the certainty that some manufacturing countries will be left out of the struggle, and that they will be available to supply one or other or both of the combatants with their Military needs ?

EAST PRUSSIAN CAMPAIGN, 1914.
STRATEGICAL MAP
Positions of Deployment shown in rings.
The ADVANCE ORDERED by General Jilinsky on 13 8/14.
Samsonoff's "DEVIATION" from that ORDER.
Railway double shown ; Railways single
Scale 1/2,000,000.
20. 15. 10. 5. 0 20 40 60 80 100 MILES
WEST
PRUSSIA
EAST
DANZIG
Dirschau
Elbing
Marienburg
Pillau
Brandenburg
KÖNIGSBERG
FORTIFIED
ZONE
Tapiau
Friedland
Heilsberg
Rastenburg
Seeburg
Allenstein
Osterode
D. Eylau
Graudenz
Passenheim
Ortelsburg
Tannenberg
Neidenburg
Strasburg
Lautenburg
Soldau
Chorzele
Willenberg
Mlawa
Bromberg
Thorn
Pultusk
Novo Georgievsk
WARSAW
GUARD CORPS
1ST CORPS
POSEN
Kalisch
Lodz
VISTULA
Passarge
Alle
R. Deime
17TH CORPS
20TH CORPS
VISTULA GARRISON
ZONE OF EMPLOYMENT OF MAIN STRENGTH
II ARMY
JOHANNISBURG
Ostrolenka

MAP I.

JOHN BARTHOLOMEW & SON, LTD. EDIN.

EAST PRUSSIAN CAMPAIGN, 1914. MAP OF THE GENERAL THEATRE
INTO EAST PRUSSIA IN AUGUST OF

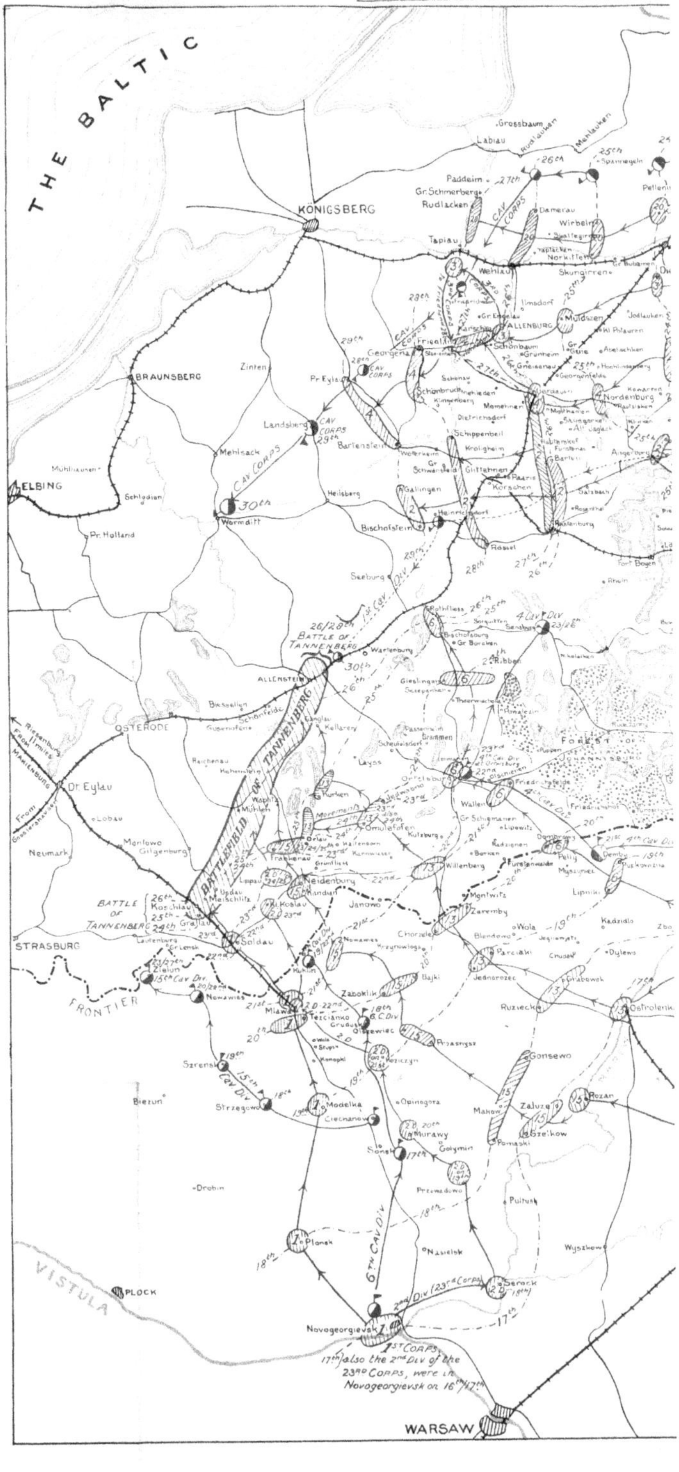

OF OPERATIONS ILLUSTRATING THE ADVANCE OF THE RUSSIAN ARMIES

THAT YEAR. Note:- The Concentration Areas are as laid down in "THE PLAN OF 1910". Map II

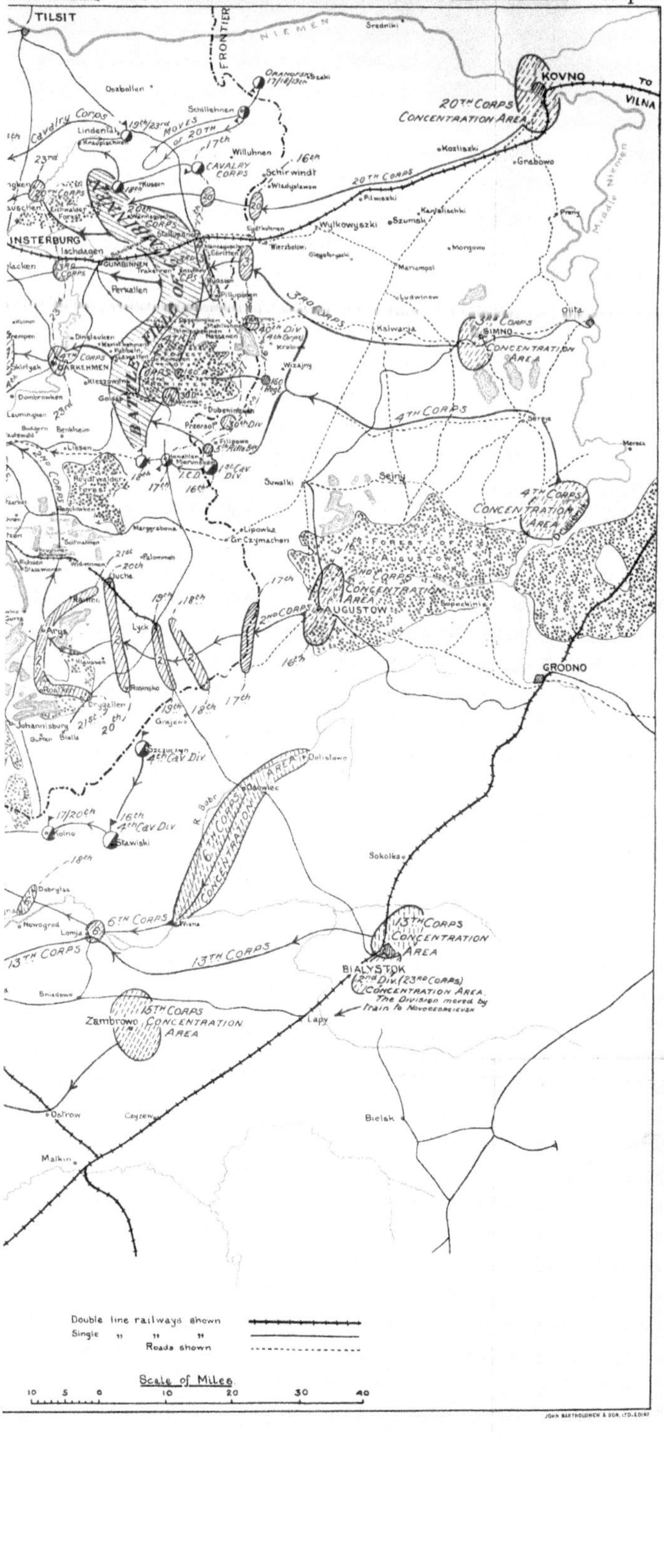

EAST PRUSSIAN CAMPAIGN, 1914.

MAP ILLUSTRATING THE A

DURING THE PERIOD 2ND

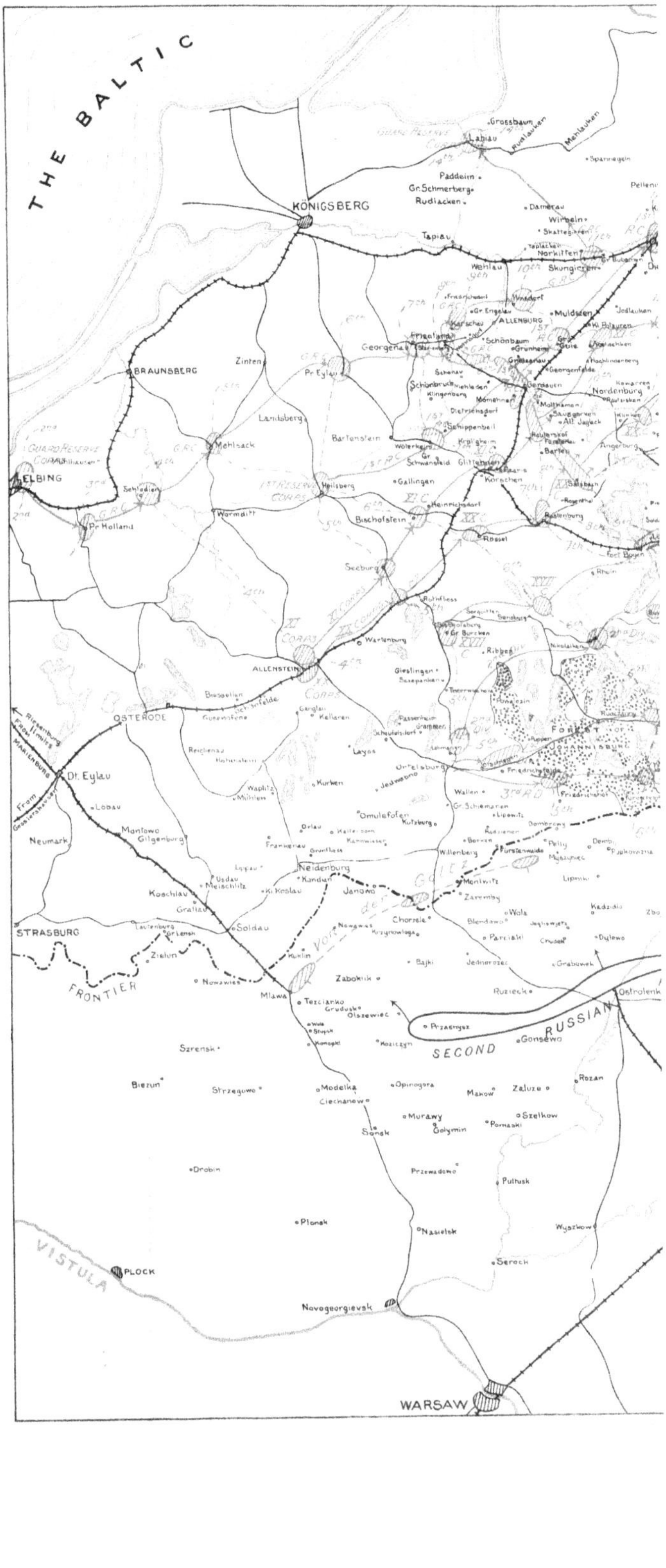

DVANCE OF THE GERMAN ARMIES TO THE 14TH SEPTEMBER, 1914.

Map III

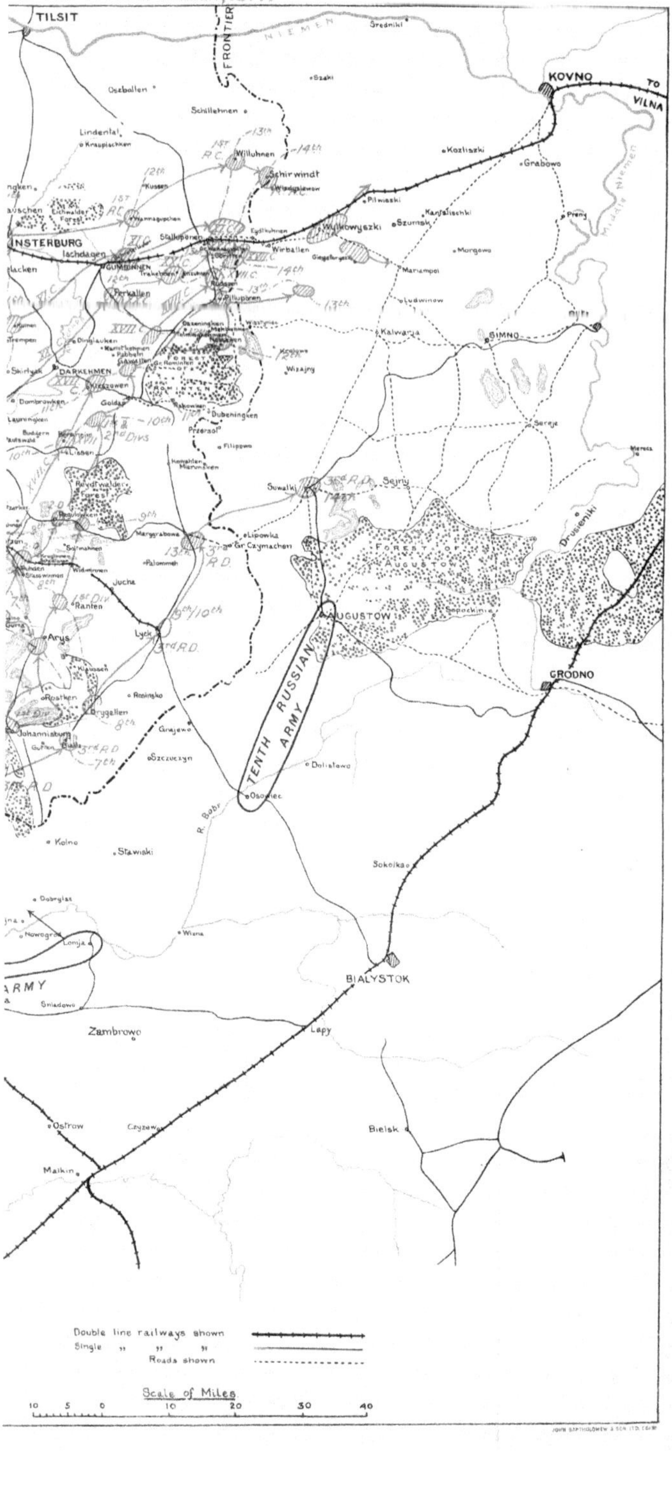

SKETCH A.

GERMAN FLANK ATTACK ON IIIRD RUSSIAN CORPS, 17TH, AUGUST.

GERMAN FRONTIER PROTECTION.

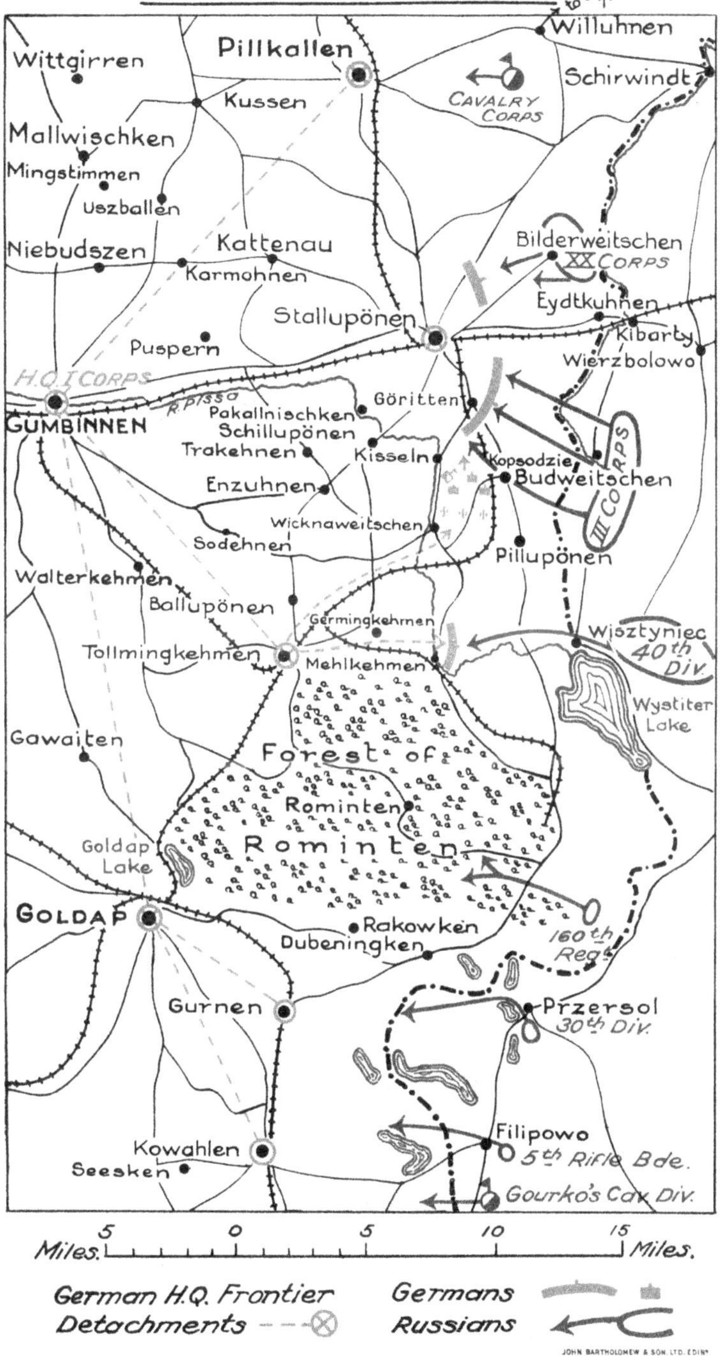

SKETCH B.

ACTION OF THE RUSSIAN 28TH DIV 19TH AUGUST.

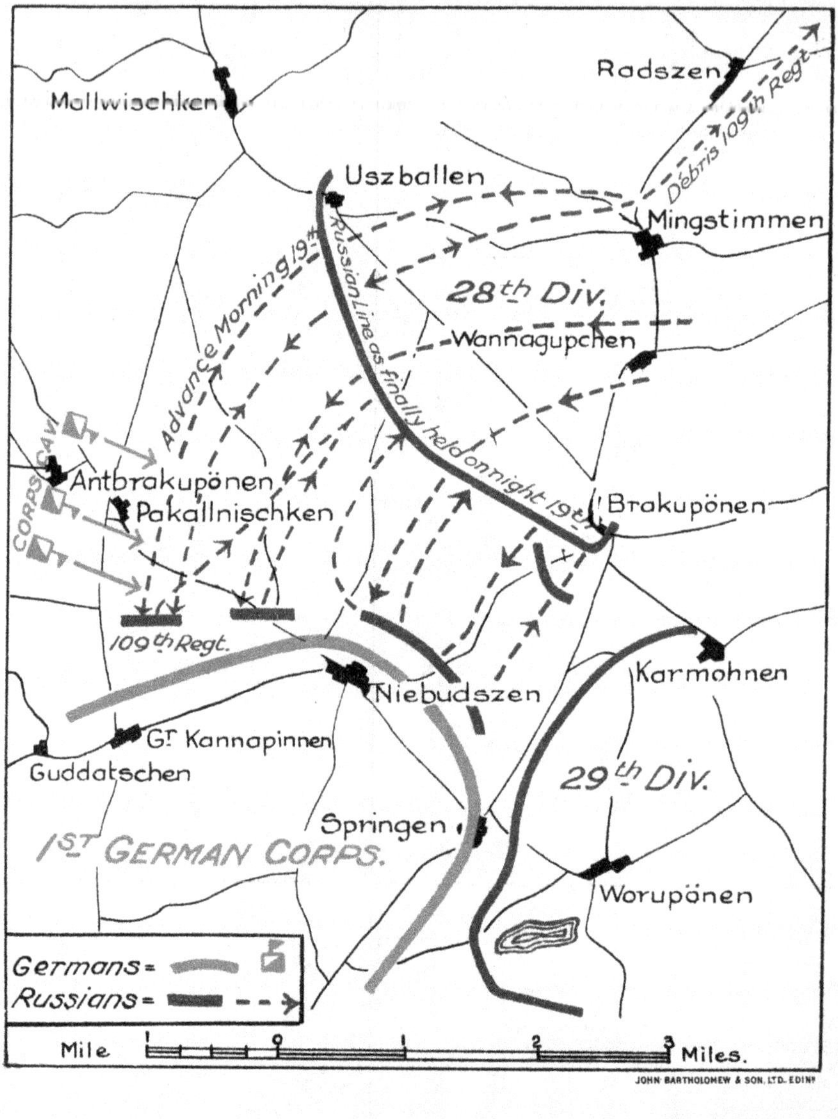

BATTLE OF GUMBINNEN.
20TH AUGUST, 1914.

SKETCH C.

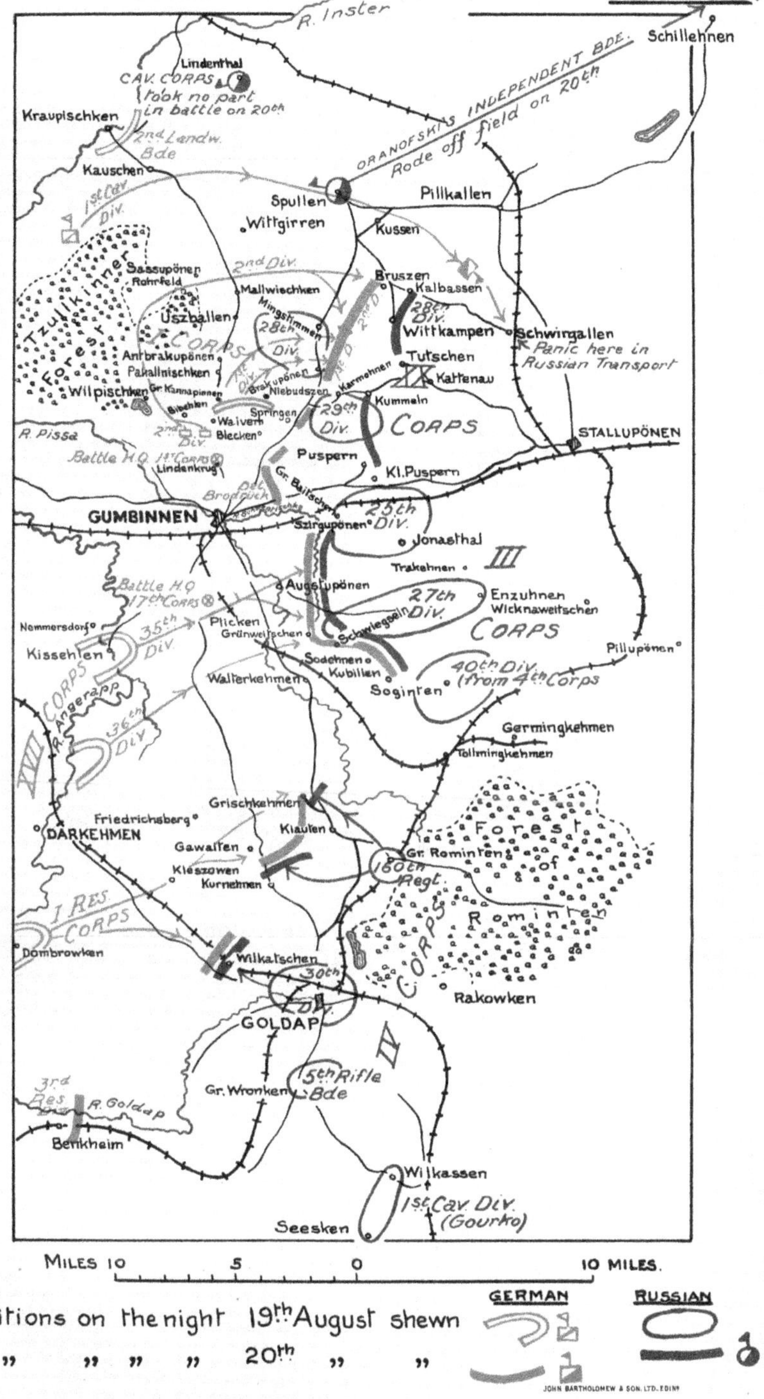

SKETCH D.

BATTLE OF ORLAU – FRANKENAU.

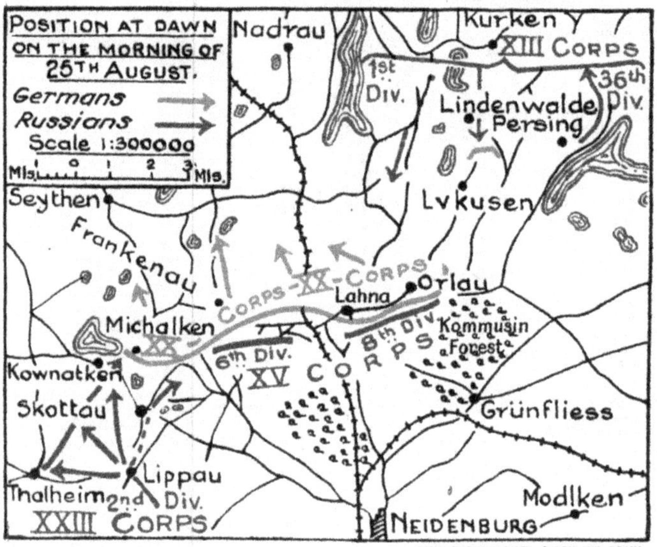

JOHN BARTHOLOMEW & SON. LTD. EDINR.

SKETCH E.

DESTRUCTION OF THE RUSSIAN RIGHT FLANK.

26TH AUG: 1914.

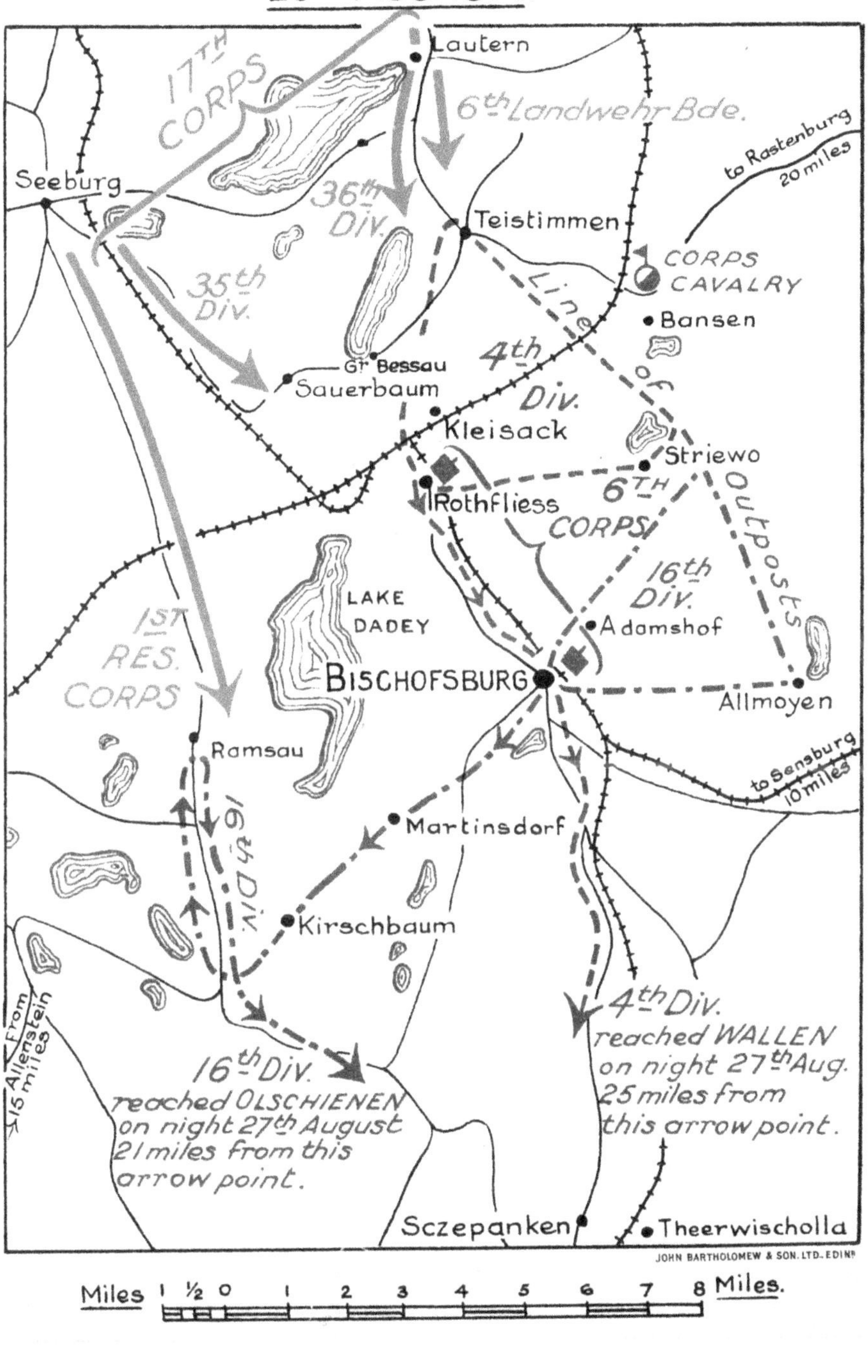

ATTACK ON RUSSIAN LI

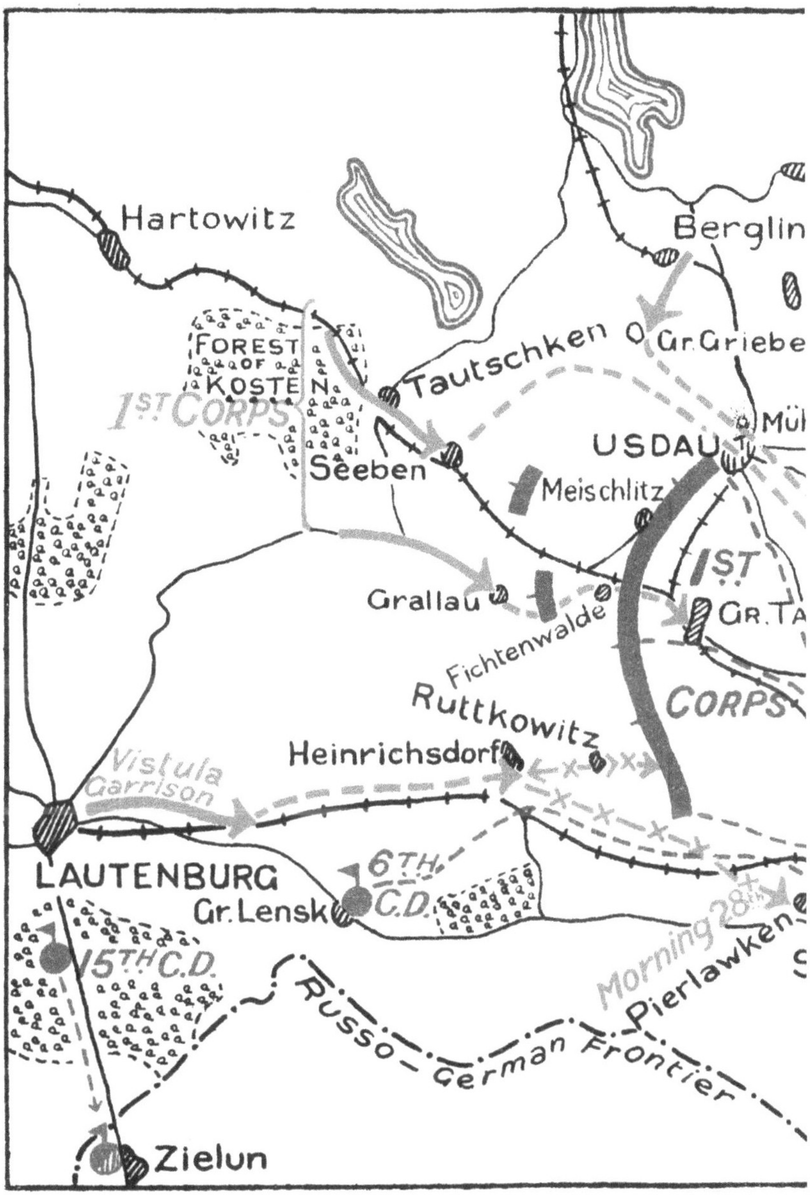

SCALE OF MILES

0 1 2 3 4 5 6 7 8 9 10

POSITIO

MOVES

MOVES

EFT, 26/28TH AUG: 1914. SKETCH F.

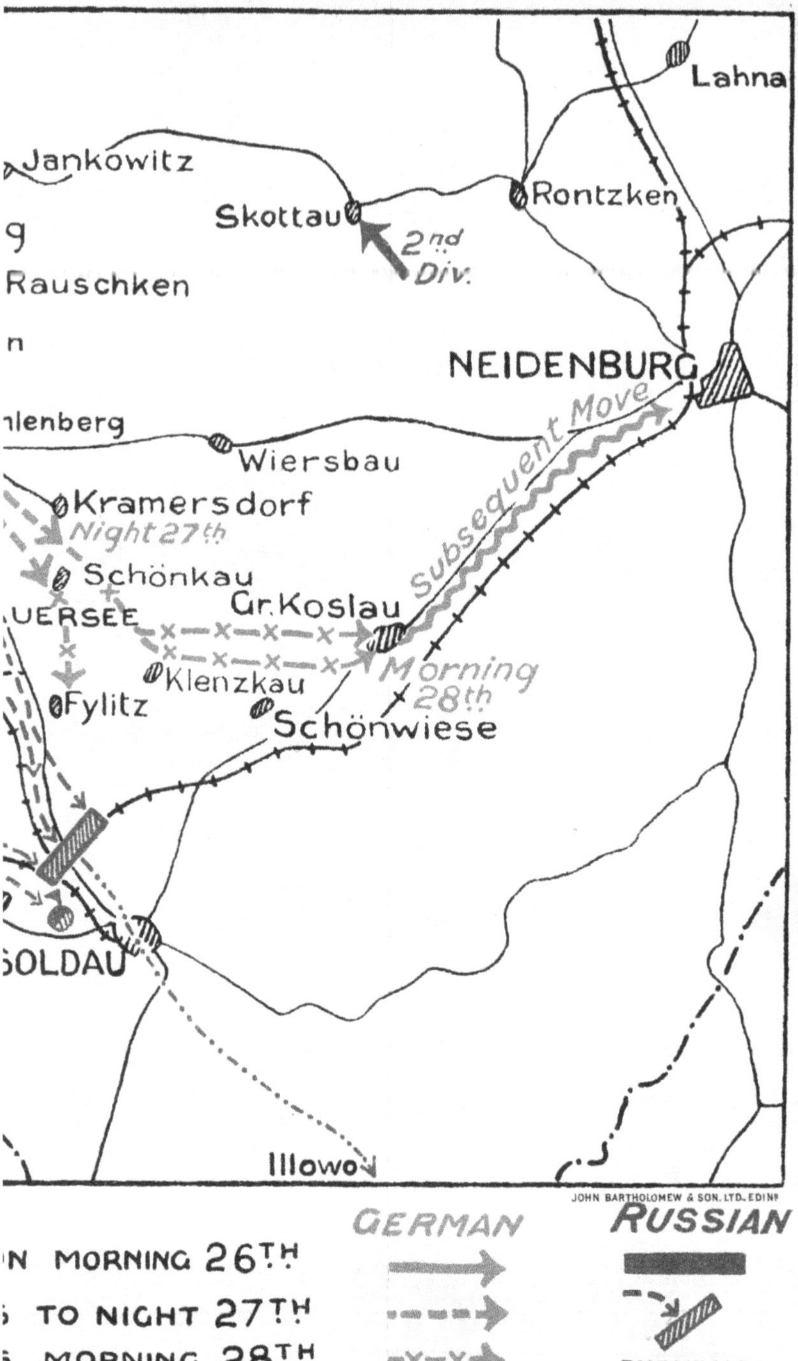

ATTACK ON RUSSIAN CENTRE 26

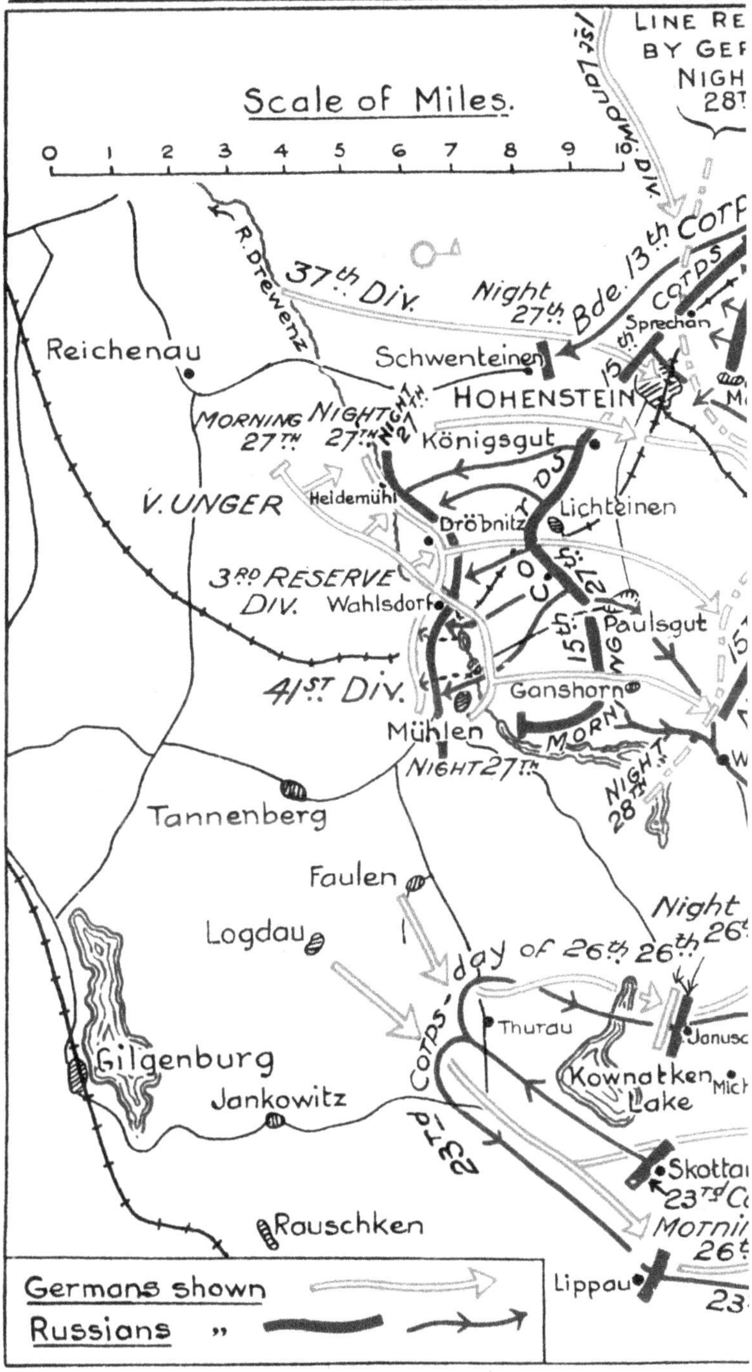

6/29TH AUG., 1914. SKETCH G.

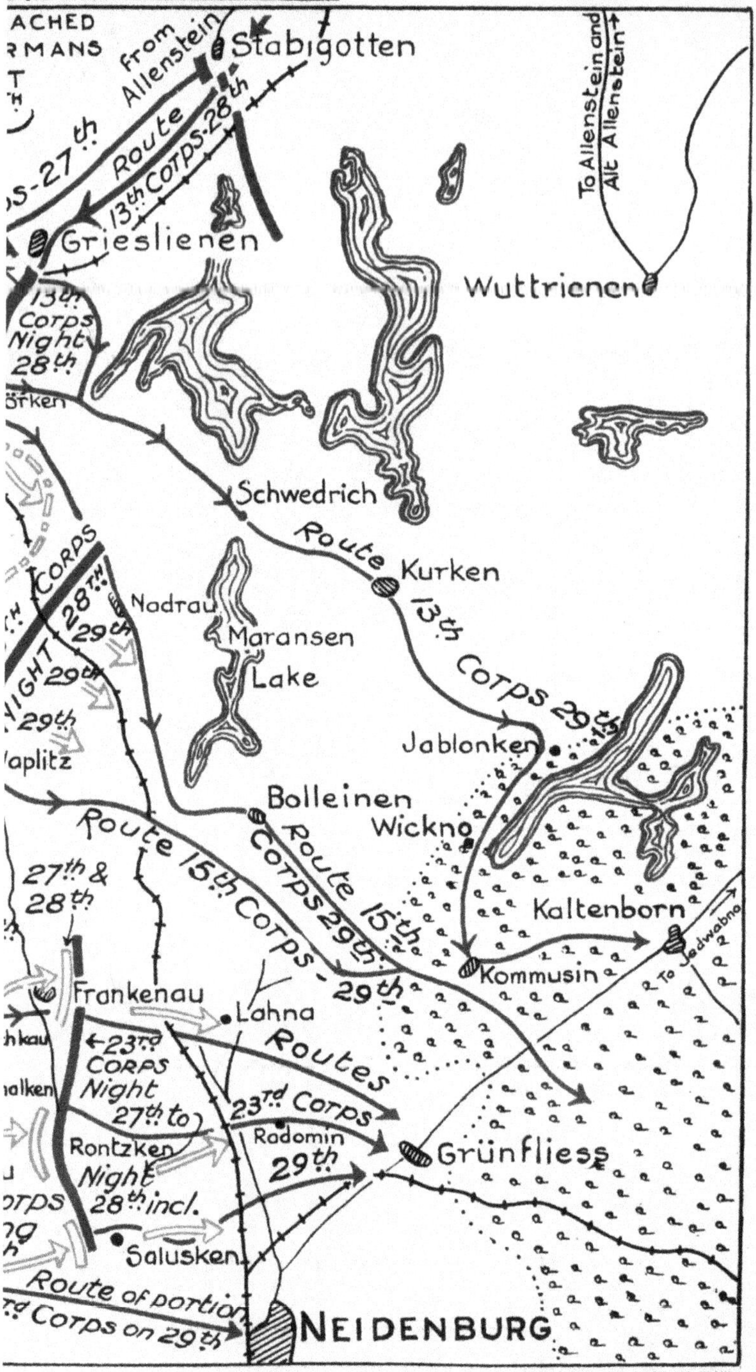

JOHN BARTHOLOMEW & SON, LTD., EDINR

THE FINAL DESTRUCTION OF

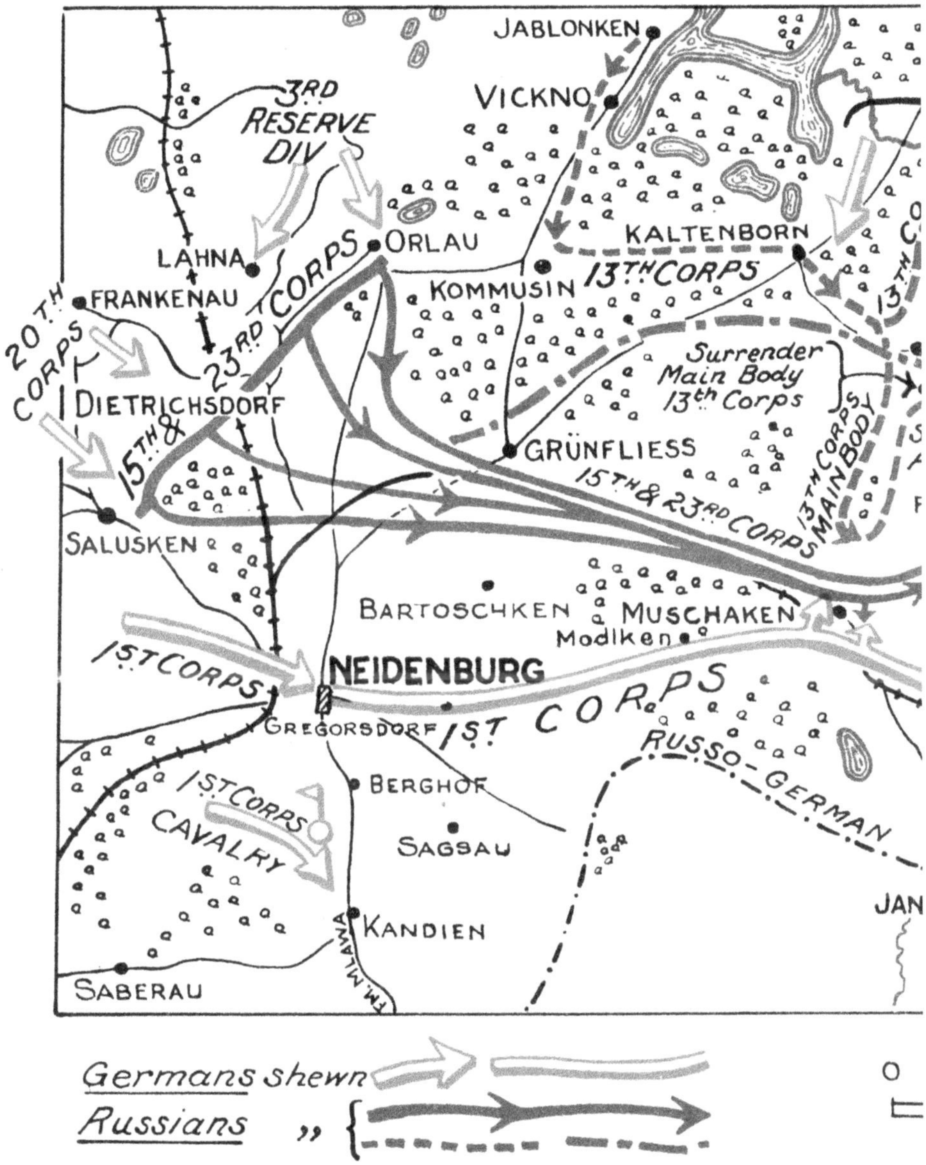

SKETCH H.

THE 2ND ARMY, 29TH-30TH AUG. 1914.

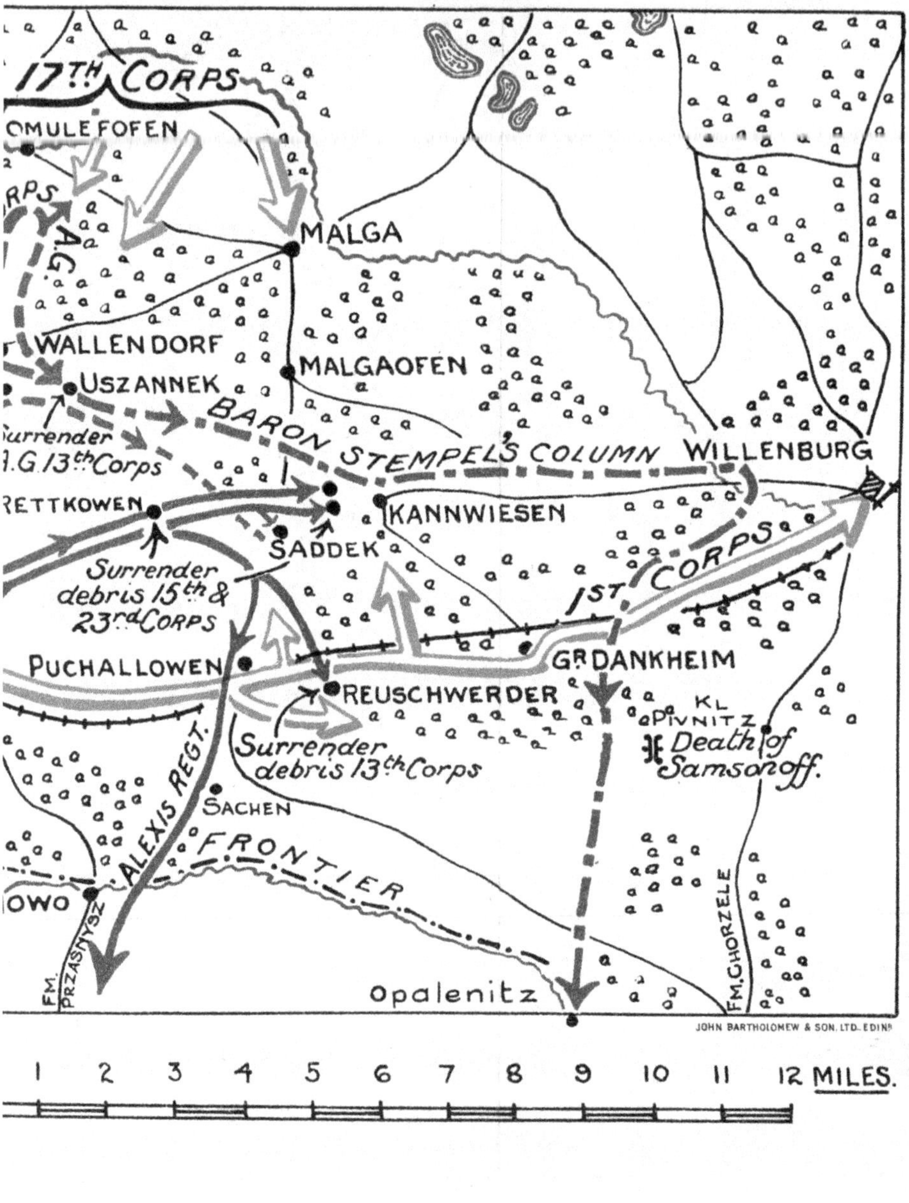

SKETCH J.

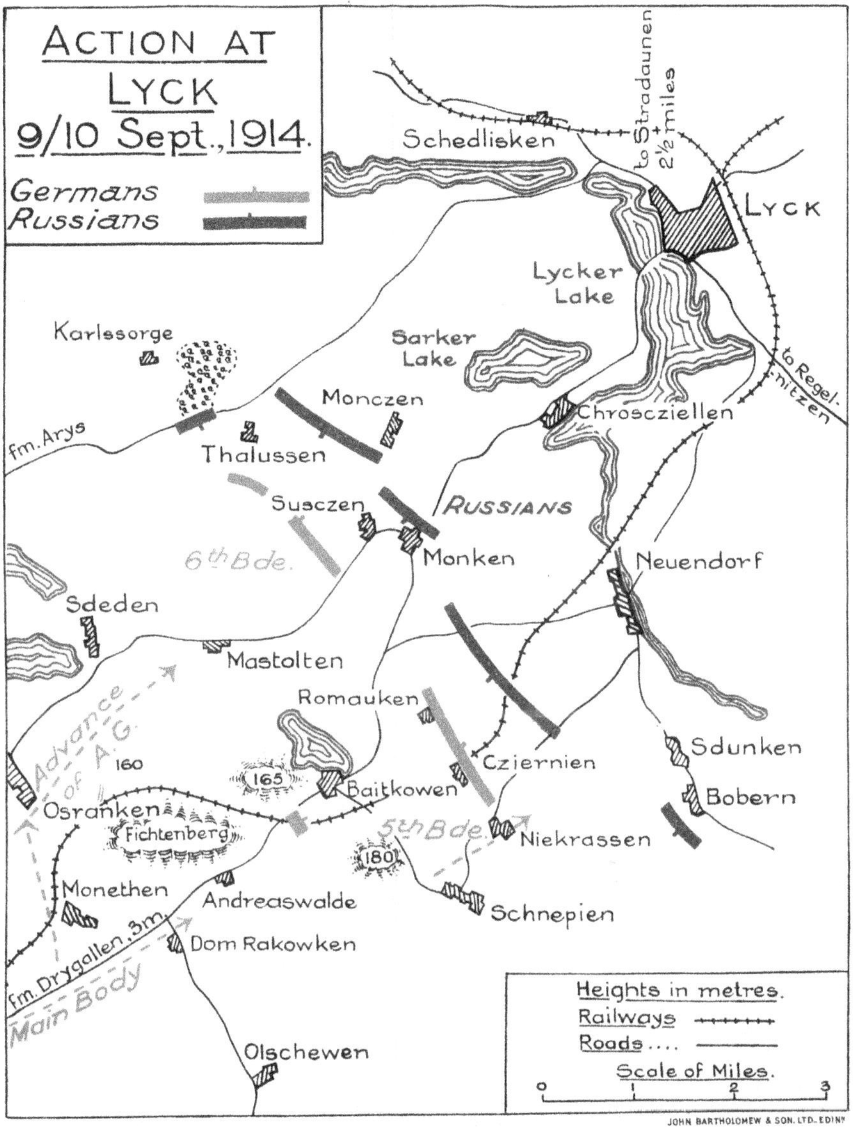

JOHN BARTHOLOMEW & SON. LTD. EDIN.

ACTION AT SOLTMAH

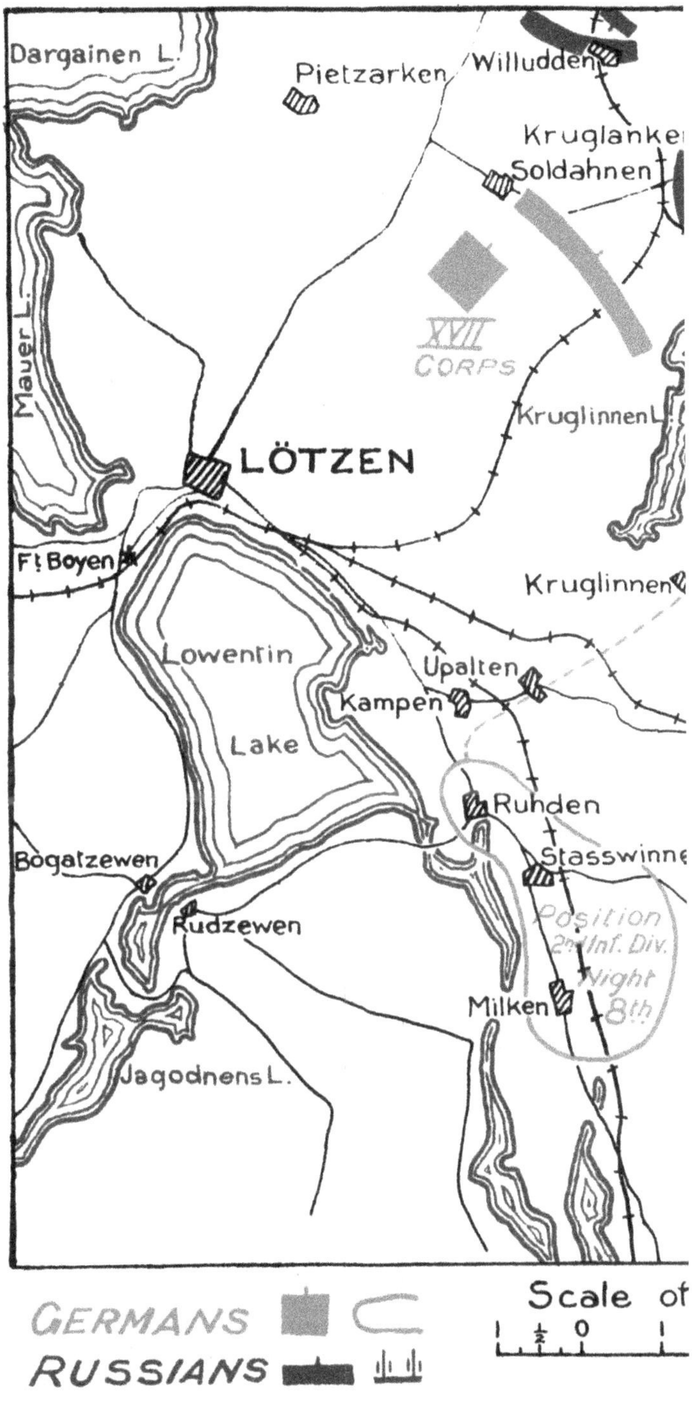

SKETCH K.

INEN 9TH SEPT. 1914.

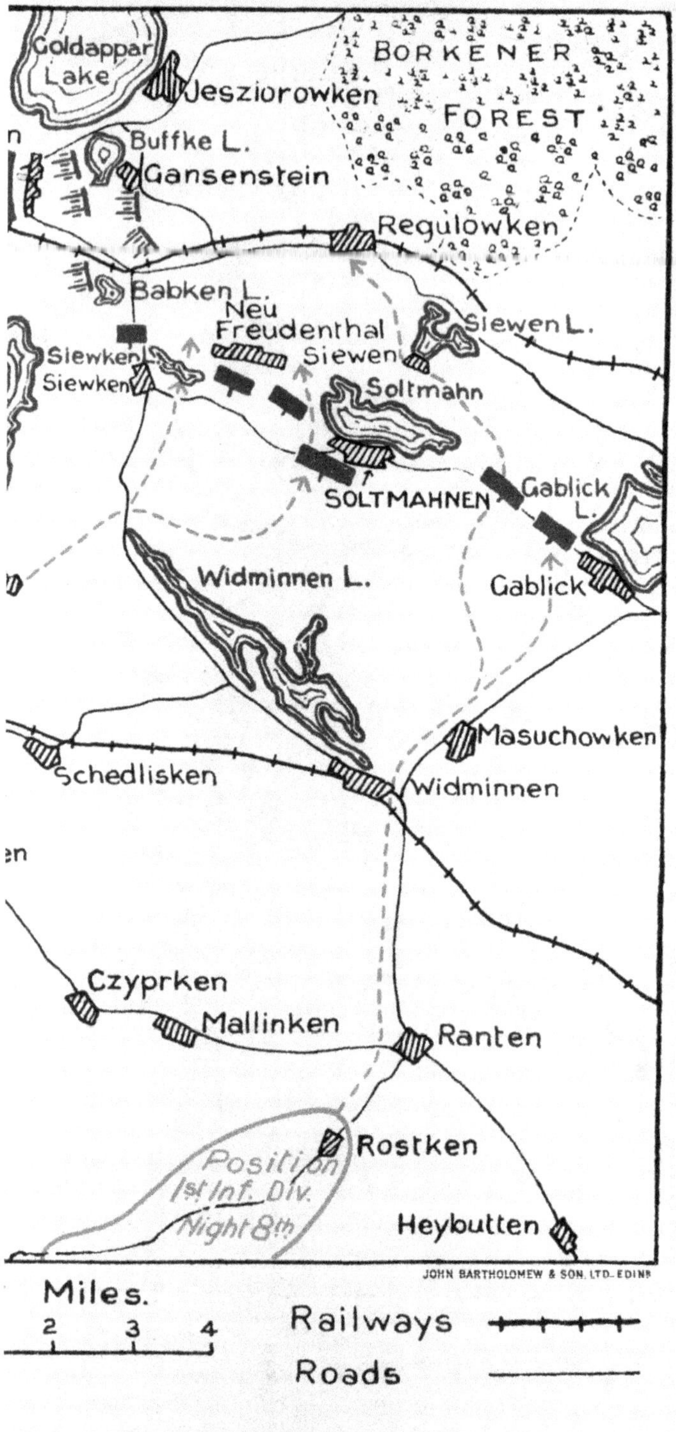

SKETCH L.

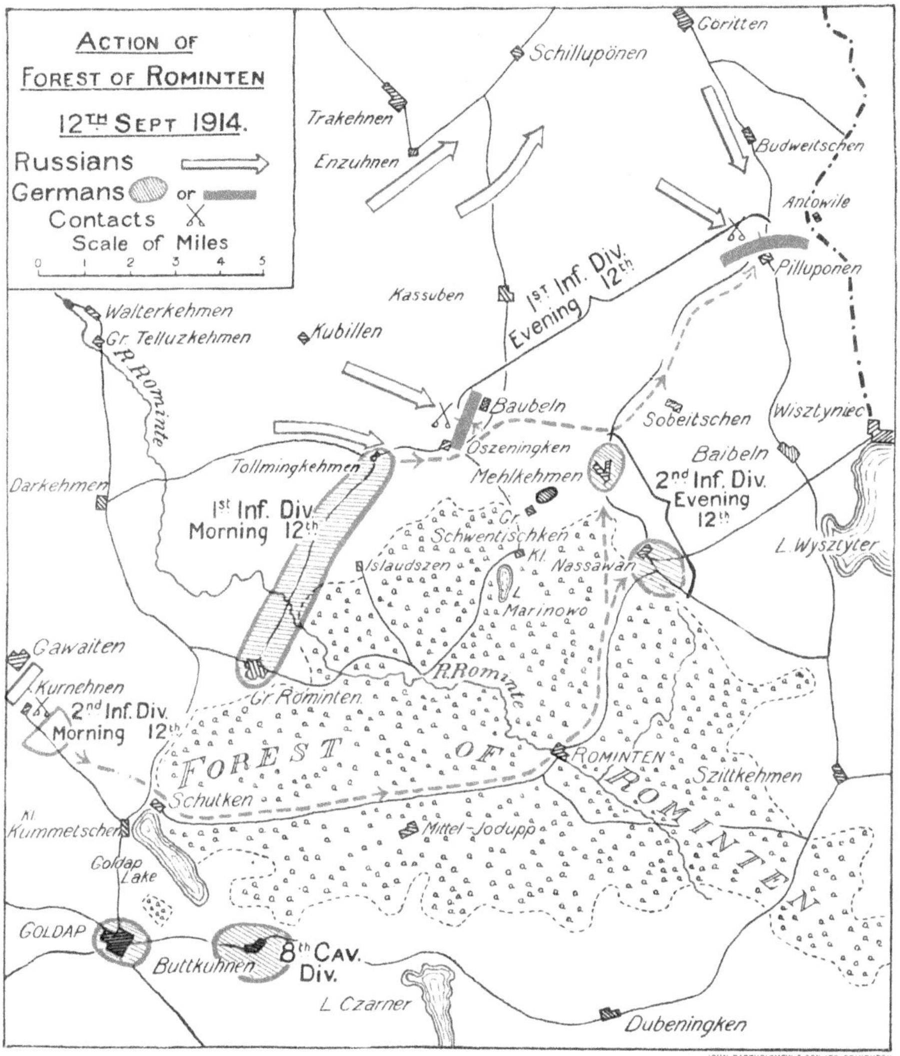

JOHN BARTHOLOMEW & SON LTD. EDINBURGH

BOOKS OF REFERENCE ON THE TANNENBERG CAMPAIGN.

ENGLISH.

1. With the Russian Army. 1914-1917. Major-General Sir Alfred Knox, K.C.B., C.M.G. Hutchinson, 1921.
2. My War Memories. 1914-1918. General Ludendorff. Hutchinson, 1920.
3. Russia in 1914-1917. General Gourko. John Murray, 1918.

GERMAN.

1. Marneschlacht und Tannenberg. General von François. August Scherl, Berlin.
2. Meiner Truppen Heldenkämpfe. General von Morgen. E. S. Mittler & Son, Berlin, 1920.
3. Tannenberg. Hermann von Giehrl. E. S. Mittler & Son, Berlin, 1923.
4. Graf von Schlieffen und der Weltkrieg. Part 1. Wolfgang Foerster. E. S. Mittler & Son, Berlin, 1921.
5. Mit Hindenburg bei Tannenberg. W. von Stephani. R. Eisenschmidt, Berlin, 1919.

6. Die Geographie des östlichen Kriegsschauplatzes. Professor Dr Tiessen. Concordia Verlag, Berlin, 1914.

7. Der Weltkrieg im Urteil unserer Feinde. General von Kuhl. E. S. Mittler & Son, Berlin. 1922.

8. Der Krieg zur Ostzee. Band 1. Marine Archiv. E. S. Mittler & Son, Berlin, 1921.

9. Der Krieg der Versäumten Gelegenheiten. General Hoffmann. Verlag fur Kulturpolitik, Munchen, 1923.

10. Erinnerungen. Briefe. Dokumente. General Helmuth von Molkte. Der Kommende Tag Verlag, Stuttgart, 1922.

11. Bis zur Marne 1914. General von Tappen. Verlag Gerhard Stalling, Berlin, 1920.

12. Entwickelung der Taktik. General Balck. R. Eisenschmidt, Berlin, 1922.

13. Der Feldherr Psychologos. Kurt Hesse. E. S. Mittler & Son, Berlin, 1922.

14. Lötzen und die grossen Masurischen Seen im Weltkrieg. Professor Ed. Schmidt. Verlag Otto Kummel, Königsberg.

15. Erinnerungen. W. A. Suchomlinow. Verlag Hobbing, Berlin, 1924.

16. Taktische Grundlehren und Leitfaden. Tannenberg. V. Stephani. Holtgebaum & Heinicke, 1924.

17. Der Weltkrieg, 1914 bis 1918. Vol. II. Die Befreiung Ostpreussens Reichsarchiv. E. S. Mittler & Son, Berlin, 1924.

RUSSIAN.

1. Краткій Стратегическій Очеркъ Войны 1914-1918 г
Русскій Фронтъ. Выпускъ 1.

2. Отчего 1 русская Армія Ренненкамфа въ августѣ 1914 г. не помогла 2 русской Арміи Самсонова.
Генералъ Радусъ-Зенковичъ.

3. Военныя соглашенія Россіи съ иностранными государствами до войны.
Генералъ Валентиновъ.

4. Стратегическій очеркъ Войны 1914-1918.
Часть 1. Я. К. Циховичъ.

5. Нѣкоторые выводы изъ сраженія при Губминенѣ. въ августѣ 1914 г.

6. Операція 2 арміи въ В. Пруссіи въ августѣ 1914 г.

www.ingramcontent.com/pod-product-compliance
Lightning Source LLC
LaVergne TN
LVHW062114200926
848484LV00007B/1670
* 9 7 8 1 8 4 7 3 4 0 7 6 4 *